A Structural Analysis of Complex Aerial Photographs

ADVANCED APPLICATIONS IN PATTERN RECOGNITION
General editor: Morton Nadler

A STRUCTURAL ANALYSIS OF COMPLEX AERIAL PHOTOGRAPHS
Makoto Nagao and Takashi Matsuyama

PATTERN RECOGNITION WITH FUZZY OBJECTIVE FUNCTION
ALGORITHMS
James C. Bezdek

A Structural Analysis of Complex Aerial Photographs

Makoto Nagao
and
Takashi Matsuyama

Kyoto University
Kyoto, Japan

SPRINGER SCIENCE+BUSINESS MEDIA, LLC

Library of Congress Cataloging in Publication Data

Nagao, Makoto, 1936-
  A Structural Analysis of Complex Aerial Photographs.

  (Advanced applications in pattern recognition)
  Bibliography: p.
  Includes index.
  1. Image processing. I. Matsuyama, Takashi. II. Title. III. Series.
TA 1632.N27   621.36'7   80-23322

ISBN 978-1-4615-8296-0   ISBN 978-1-4615-8294-6 (eBook)
DOI 10.1007/978-1-4615-8294-6

© Springer Science+Business Media New York 1980
Originally published by Plenum Press, New York in 1980
Softcover reprint of the hardcover 1st edition 1980

FOREWORD

It is most appropriate that the first volume to appear in the
series "Advanced Applications in Pattern Recognition" should be this
monograph by Nagao and Matsuyama. The work described here is a deep
unification and synthesis of the two fundamental approaches to pattern recognition: numerical (also known as "statistical") and structural ("linguistic," "syntactic"). The power and unity of the methodology flow from the apparently effortless and natural use of the
knowledge-base framework illuminated by the best results of artificial
intelligence research.

An integral part of the work is the algorithmic solution of many
hitherto incompletely or clumsily treated problems.

It was on the occasion of a laboratory visit in connection with
the 4th IJCPR (of which Professor Nagao was the very able Program
Chairman) that I saw in operation the system described here. On the
spot I expressed the desire to see the work described for the international technical audience in this series and the authors were kind
enough to agree to contribute to a new and unknown series.

With the publication of this monograph on the eve of the 5th ICPR
my wish is fulfilled. I want to thank here the authors and Plenum
Publishing Corporation for making this volume and the series a reality.

Morton Nadler

La-Celle-St-Cloud
August 15th, 1980

PREFACE

This book is a result of our research activities in image analysis and recognition during the past few years. The main subject presented in this book is the structural analysis of aerial photographs, of suburban areas in particular, which show very complex geographical structures. The analysis system automatically locates a variety of objects in an aerial photograph by using diverse knowledge of the world. It is one of the first image understanding systems that has incorporated very sophisticated artificial intelligence techniques into the analysis of complex natural scenes.

The whole system of image analysis is built on the basis of the concept of a production system. It has several excellent features, such as modularity of picture processing subcomponents, flexibility of augmenting analysis procedures, heterarchical control structure, and facilities for error correction and backtracking.

The book also presents many new picture processing techniques, such as edge-preserving smoothing, structural texture analysis, longest skeleton detection, measure of elongatedness, valley detection in various histograms, and so on, which have been successfully applied to the analysis of aerial photographs.

The research is deeply involved in a wide variety of topics in the areas of pattern recognition, picture processing, remote sensing, and knowledge representation and control structure problems in artifical intelligence. The book, therefore, will be interesting and valuable to those who are engaged in these research areas.

The major parts of this research were performed as the Ph.D. thesis of T. Matsuyama under the guidance of Prof. M. Nagao of Kyoto University, Japan.

Some programs and ideas presented in this book were published in the proceedings and journals listed below. According to the recommendation of Dr. M. Nadler, the general editor of this book series, we have summarized the whole scope of our system into the book.

We wish to thank Dr. T. Kasvand of National Research Council, Canada, for his enduring efforts of reading through the manuscript and of correcting errors in English. We also express our thanks to the members of our laboratory for their collaboration in the implementation of the system. Finally, we are grateful to Miss Y. Yamamoto for her typing of the draft of the manuscript.

Related papers by the authors:

(1) M. Nagao and T. Matsuyama, "Edge Preserving Smoothing", *Computer Graphics and Image Processing,* Vol. **10**, 1979.

(2) M. Nagao, T. Matsuyama and Y. Ikeda, "Region Extraction and Shape Analysis in Aerial Photographs", *Computer Graphics and Image Processing,* Vol. **10**, 1979.

(3) M. Nagao, T. Matsuyama and H. Mori, "Structural Analysis of Complex Aerial Photographs", Proc. of 6th IJCAI, Tokyo, 1979.

(4) T. Matsuyama and M. Nagao, "Structural Analysis of Aerial Photographs", *Jour. of IPS Japan*, 1980 (in Japanese).

CONTENTS

LIST OF TABLES

LIST OF FIGURES

[1] The color insert follows page 104.

1. INTRODUCTION

1.1. Preliminary Remarks

Studies of pattern recognition and picture processing have made great advances in the last few decades, and their fields of application are spreading out more and more widely, being encouraged by the rapid progress in computer hardware and LSI technologies.

A variety of theoretical and practical methods have been developed to process digital pictures by computer. (Recently several books on digital picture processing have been published [22,26,59,64]. For general methods and ideas of digital picture processing, refer to these books.) Generally speaking, the methods in digital picture processing can be classified into two different categories according to the objective of the processing. In the first case, an input picture is transformed into a new picture so as to improve the pictorial information for human interpretation. Smoothing, edge and line detection, pseudo-color representation, and digital filtering are the most popular techniques to enhance the quality of the picture and to reduce the degradation by noise. These methods are called enhancement and restoration, and have been extensively studied in connection with the space project in the United States. On the other hand, there have been strong desires to automate various tasks which require human pattern recognition abilities. For these purposes, various feature extraction and classification methods have been developed to bestow the ability of visual perception on the computer. Shape and texture analysis is one of the most active areas in picture processing. Character recognition has the longest history among various applications, and some commercial machines for automatic character recognition have been put into practical use. Recently, the research called image understanding and scene analysis has come to be widely studied. Its aim is to describe the scene by analyzing an array of sensory data with the help of knowledge about the scene.

With the development of theoretical and practical procedures, more and more complex images have come to be processed. The analysis

1

of biomedical and remote-sensing images are the most active application fields. Especially in these fields, a very huge number of images are routinely produced so that there exists a strong demand to automate the processes of image analysis and quantitative measurements by using computers.

Remote sensing is the name of the technology of acquiring information about an object or a phenomenon from physical measurements made without direct contact with the object. In most cases, the measurements are made by recording the reflected electromagnetic energy in several different spectral bands. The information is represented as pictorial data. The application fields of remote sensing technology include agriculture, forestry, hydrology, geology, geography, cartography, and environment monitoring. (For a general introduction and analysis methods used in remote sensing, refer to [8,9,44,52,72].)

There are two kinds of platforms from which the measurements of the situation on the ground surface are made: satellites and airplanes. Satellites, such as LANDSAT, give very large-scale images of the earth's surface for the global investigation of the earth's resources. On the other hand, images taken from an airplane are of very high resolution and suitable for the detailed examination of the land-use patterns in local districts. Especially in urban and suburban areas, as the land-use patterns are very complex and tend to change in a short period, it is very important to be able to make immediate judgements regarding the current situation by analyzing aerial images. The detailed analysis of aerial images includes the enumeration of the number of houses, the measurement of the area of vegetation, the description of road networks, and so on. The results of such analysis could be used to make a detailed land-use map and to grasp the situation of a district in order to facilitate future planning.

The LANDSAT project accelerated the progress of various techniques in digital picture processing. Those for registration, geometric and radiometric correction, and enhancement have been extensively studied, and several standard methods have been established.

On the other hand, as the current methods of visual inspection by human photointerpreters are often very time-consuming and costly, it has been strongly desired and would be quite valuable to automate the process of interpreting aerial photographs. So far various methods have been proposed for the classification and detection of objects in remotely sensed imagery. These methods, however, are too simple for describing the structures in complex aerial photographs of urban and suburban districts. Recently, several systems have been developed for interpreting the situations on the ground surface by using artificial intelligence techniques for scene analysis and image understanding.

The system presented in this monograph has been developed to locate various kinds of objects in an aerial photograph and to obtain a description of the structure on the ground surface. It is an image understanding system where the diverse knowledge is incorporated to recognize objects and describe the scene. All the processes are performed automatically without any help of human researchers, and we can get quantitative measurements of objects on the ground, such as locations, numbers, sizes, etc.

In this chapter, we will briefly review the methods of ordinary statistical classification and target detection used to analyze remotely sensed imagery, and will point out several intrinsic short-comings in these methods. Then, the ideas of structural analysis in our system will be presented in connection with the research on scene analysis and image understanding.

1.2. Statistical Classification Methods

In the statistical classification methods, each sample is char-acterized by a set of features, and the classification of the samples is made by dividing the feature space, based on the distributions of the feature vectors of the samples.

The remote sensing data are usually taken in several different spectral bands. Thus, each point in a picture can be characterized by the feature vector whose components correspond to the intensity levels in each spectral band. The most straightforward way of applying pattern recognition methods to the remote sensing problem is to consider each point in a picture as a sample and to classify it on the basis of its spectral properties. The maximum-likelihood method, linear discriminant functions, and clusterings are the most popular methods for classification. (For theoretical discussions on these classification methods, refer to [18]. Fu [21] surveyed various applications of pattern recognition techniques to remote sensing images.)

These statistical classification methods have been quite widely used for various applications, such as classification of crop field plantations, being supported by well-defined theoretical backgrounds. These methods, however, rely only on the multispectral characteris-tics of a point without considering the spatial relations with its neighboring points. That is, the statistical models used in these classification methods characterize the distributions of samples in the feature space but disregard the spatial information in the two-dimensional picture space. As a matter of course, these simple methods can not give satisfactory results in analysis of complex aerial photographs. They have several intrinsic shortcomings, such as:

1. They are very sensitive to random noise.

2. They cannot handle areas with heavy texture

3. As aerial photographs are taken under different conditions, the reference spectral property of each category,
   taken from one picture, cannot be directly applied to
   the classification of other pictures.

4. They cannot discriminate different objects with similar
   spectral properties because they rely only on multispectral characteristics.

5. The classification results can not give the information about the numbers and shapes of the objects in
   a scene.

These problems have been pointed out at the very early stages
of the research on remote sensing, and various refined methods which
incorporated digital picture processing techniques have been proposed.

There have been several efforts to incorporate spatial information into the classification in order to overcome noise and texture.
Nagao *et al.* [50] utilized an intensity histogram in a small area
centered at each point to characterize the textural property of the
area, and classified each point on the basis of the similarity
between intensity histograms. Haralick *et al.* [29], Tamura *et al.*
[73], and Weszka *et al.* [81] used several textural features calculated from the "gray-level co-occurrence matrix" [29] to classify
crop field plantations and terrain patterns. In these methods, each
point in a picture is characterized by some statistics calculated
from a set of gray levels of its neighboring points, so that the
results are insensitive to noise and texture or, rather, these
methods discriminate points according to textural properties.

Huang [36] proposed a per-field classification method where
first a picture was segmented into regions and then classification
was made by taking each segmented region as a unit. This method is
based on the idea that if the majority of the points in a region are
classified as a certain category, then it seems reasonably sure that
the entire region as a whole belongs to that category. As each point
in a region is classified into the same category, the result becomes
very smooth and free from noise. The problem here is the determination of regions. Ketting and Landgrebe [41] and Pavlidis and
Horowitz [56] proposed segmentation methods for remote sensing images.

Fukada [23], Haralick *et al.* [28,30], and Nagy *et al.* [53]
proposed spatial clustering procedures which incorporated into clustering the spatial relationships among points. They first group
neighboring points with similar intensity (color) into homogeneous
regions, and then perform clustering in the feature space taking each
region as a sample.

In order to attain stable analysis despite the changeable photographic conditions, a method called "signature extension" has been
developed. In the supervised classification methods such as the
maximum-likelihood method, whenever we want to analyze a new picture,

we must specify training areas since the multispectral properties of
objects are apt to change from picture to picture. Henderson [32]
used a linear transformation to correct the differences in gray
levels between two pictures. Nagao *et al.* [51] proposed a classification method based on the ordering relation among the average gray
levels of regions. In texture analysis a picture is usually normal-
ized to have a certain standard gray level distribution in order to
evade effects of changeable illuminations.

Although these methods have certainly achieved some improvements
in overcoming the shortcomings 1 − 3, they cannot go beyond the
limitations of the classification. That is, the results of these
statistical classification methods show, in essence, to which category each point in a picture belongs. They cannot give information
about the objects in a scene. For example, roads and buildings
sometimes have the same spectral properties because they are made of
concrete. In this case, the statistical classification methods have
no way of discriminating these two objects. Moreover, even if all
objects in a scene had different spectral properties, it would be
impossible to obtain quantitative measurements related to the objects
such as the number of buildings.

Thus, it might be true to say that ordinary statistical classification methods do not recognize objects but just classify points or
regions, and that the classification itself is unstable and dependent
on the picture under analysis. Their most serious shortcoming is the
inability to incorporate the concept of "object" into the processing;
they do not utilize the available knowledge about the various properties of objects such as shape, size, location, syntactic and semantic
relationships with other objects, and so on.

1.3. Target Detection

Target (object) detection is the name for methods to locate
specific objects in a picture using the knowledge of their distinguishing properties such as color, shape, size, etc. Research on
target detection by computer started as early as 1960, and advanced
in parallel with the development of statistical classification
methods.

In the early stage of this research [16,33], the locations of
objects in a picture were often detected by "template matching"
techniques, which had been very popular methods in optical picture
processing. A template of an ideal object is scanned systematically
across a picture, and the similarity between the template and the
picture is calculated at each position to determine whether or not
the object is present at that position. Since the size and direction
of the object is not known in advance, one has to prepare a set of
templates of different sizes and directions for detecting the objects
in various situations. Scanning with multiple templates requires

much processing time. Moreover, this simple technique can not be made
adaptive to unexpected changes in the shape of an object. Therefore,
the objects to be detected by template matching techniques are restricted to those of a fixed form.

Digital picture processing techniques such as edge and line
detection and curve tracking made it possible to extract such
variable-shaped objects as roads. Bajcsy and Tavakoli [5], and Li
and Fu [43] extracted line segments from satellite images to locate
roads in urban and suburban areas. Barrow and Fischler [7] used a
sophisticated tracking technique to detect roads in aerial photographs.

Recently, many "expert programs" have been developed which
utilize some knowledge about the structures of objects to locate, for
example, various vehicles [1,13,47-49,60]. Since it is very difficult to detect directly an object with a complex structure, they first
extract basic components of the object by picture processing techniques, and then locate the object considering spatial relationships
among these components. That is, they utilize the knowledge of the
structures of objects.

These target detection programs dichotomize the visual world
into objects and background, and utilize visual models of objects as
the knowledge source for object recognition. However, in the analysis
of complex aerial photographs, in order to obtain the descriptions of
the situation on the ground surface, these ordinary target detection
techniques do not work well if we rely only on the knowledge of the
object itself. There are various context-sensitive objects whose
detection requires information about the environment of the objects
in which they are embedded. For example, it is very difficult to
detect cars by the simple knowledge that they are rectangular, without knowing that they are on roads or in parking lots There are many
other objects which look rectangular when seen from above. Therefore,
the detection of the cars will entail the recognition of roads and
parking lots. There are many other examples of this kind. Ships will
be found on the water surface, rivers may have bridges, houses may
have roads and gardens adjacent to them, and so on. Environmental
knowledge of this kind makes the detection of these objects both easy
and reliable. Thus, the syntactic and semantic constraints among
objects as well as the intrinsic properties of objects are essential
to the automatic photointerpretation of complex aerial photographs.

1.4. Image Understanding Applied to Aerial Photographs

Research on image understanding (scene analysis) and computer
vision is one of the most challenging fields in artificial intelligence. Its aim is to find out objects in a scene and to describe
their structures and mutual relationships by analyzing an array of
sensory data. The research started with the recognition of three-

dimensional configurations of simple blocks [80,83] and proceeded to
handle more complex scenes such as indoor scenes [24,75] and outdoor
scenes [4,68,84].

The most distinguishing characteristic of image understanding
is that it constructs a description of the scene whereas statistical
classification methods label each point in a picture with category
names and picture processing techniques transform pictures into other
pictures. As a matter of course, the processes of detecting objects
and describing the scene require both well-structured knowledge and
some sophisticated control structure. While the core of image understanding is a knowledge-based symbolic processing, a process of
organizing picture data and its interfacing with the symbolic process
are also very important for an image understanding system to analyze
complex natural scenes. (For a survey of model representations and
control structures in image understanding, refer to [39].)

In the analysis of aerial photographs, in order to overcome the
before-mentioned limitations of ordinary statistical classification
and target detection, it is natural to introduce artificial intelligence techniques used for image understanding. Indeed, in these days
several systems including ours have been developed which incorporate
diverse knowledge to locate objects in aerial photographs and to guide
the analysis process.

Figure 1.1 shows the schematic drawing of the developments in the
analysis of remote sensing images. In statistical classification,
with the development of region-based analysis, the concept of "object"
has been introduced into the analysis, and the spatial properties of
regions as well as the multispectral properties have come to be used.
In target detection, the knowledge of contextual and semantic constraints among objects has been introduced to recognize context-
sensitive objects. In this way, the analysis of remote sensing images
has come to incorporate more and more complex and diverse knowledge.
Then, artificial intelligence techniques for knowledge representation
and control mechanism have been introduced to organize such knowledge
and to realize a sophisticated analysis.

1.4.1. Knowledge Sources in the Analysis of Aerial Photographs

When we apply techniques of image understanding to aerial photographs, it is very important to pigeonhole knowledge sources useful
for the analysis. The knowledge sources for the photointerpretation
of remote sensing images may be classified into the following three
categories:

1. **Knowledge about photographic conditions.** Since the information about the district, the date, the altitude, and the weather
under which an aerial photograph is taken is given before the

analysis, we can utilize such information to guide the analysis.

Map data of the district under investigation will be quite valuable to locate objects in a complex situation. The map information can be considered as a good model of the real situation even though it may only be an approximation and out of date. When we utilize map data in the analysis of remotely sensed images, the most serious problems are how to register an image with the map and how to store the map data in a computer.

In the HAWKEYE system developed at Stanford Research Institute [6,11] a generalized digital map is used to guide the process of image interpretation. It contains a set of typical landmarks to match an image with a reference map. Once the image is registered, the system can roughly estimate from the map information areas corresponding to time-invariant objects, such as roads and harbors. Then, it examines such areas in detail to locate cars and ships.

Horn and Backman [34] and Horn [35] proposed a method to register real images with a map by using synthetic images created from the digital terrain map and the direction of the sun.

2. Knowledge about the intrinsic properties of objects. In order to discriminate objects, we have to characterize them by a set of features such as shape, size, location, color, texture, etc., by using picture processing techniques.

In remotely sensed data, since we can use the multispectral information of objects, we have more information than the usual color images of indoor and outdoor scenes. That is, we have images taken in invisible spectral bands as well as in red, green, and blue bands, so that we can estimate materials of objects more correctly. Spectral characteristics of objects, however, tend to change depending on the photographic conditions. Therefore we must find the effective and stable features of spectral characteristics of objects through physical experiments. On the other hand, spatial properties of objects such as size, shape, and texture are very stable and will become the distinguishing features for discriminating objects.

In order to calculate these features as correctly as possible, we must incorporate various picture processing techniques and measuring methods into image understanding systems for aerial photographs. It may be sometimes quite useful to quantify the knowledge which human photointerpreters use to discriminate objects. For example, in order to classify tree types in forest areas by textural properties, we should make quantitative measurements of the properties used by human specialists.

In addition to these features, which refer to points and regions in an image, we have to utilize the knowledge about the structures of objects in order to recognize complex objects composed of several different parts. Aggarwal and Wittenburg [1] recognized several tactical targets in FLIR (forward looking infrared) images by using

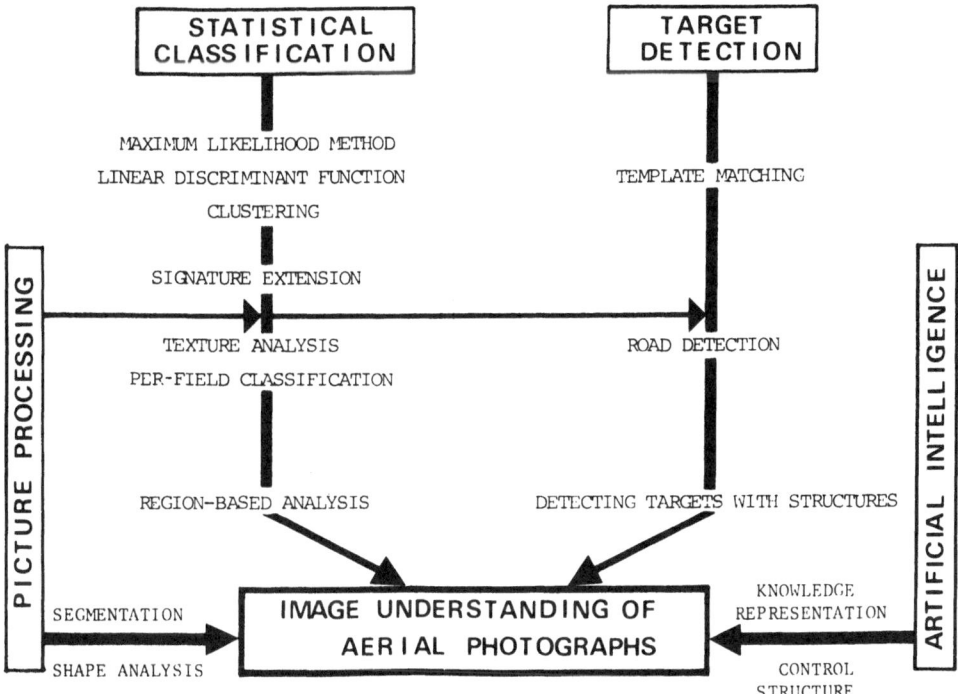

Fig. 1.1 Developments of analysis methods in remote sensing

hierarchical descriptions of their structures. The ACRONYM system at Stanford University [10] stores three-dimensional models of objects constructed from generic parts which are represented by generalized cones [54], and makes use of the observability graph to register the view angle of an aerial photograph. The more complex the structure of the objects becomes, the more important becomes the knowledge of the organization of the components.

3. Knowledge about contextual and semantic constraints among objects. As mentioned in Section 1.3, it is necessary to incorporate contextual and semantic constraints among objects when we analyze complex aerial photographs to get the detailed description of the situation on the ground surface.

Bajcsy and Tavakoli [3] and Russell and Brown [67] utilized a relational graph and a constraint network to denote the spatial relationships among objects, and they succeeded in locating bridges, islands, and aeration tanks in aerial images. The detection of these objects would have been very difficult relying only on their visual models.

When we want to recognize conceptual objects with administrative meanings such as residential, commercial, and industrial areas, we also have to incorporate the knowledge about the conceptual structures of these objects.

1.4.2. Problems in Automatic Photointerpretation

In ordinary aerial photographs, since images are taken high above the ground, the shapes of the objects do not change very much with camera position. The objects on the ground either are two-dimensional flat ones or have negligible height compared to the altitude of the camera position. Therefore, we need not worry so much about the view angle of the scene for locating the objects. In this sense the interpretation of aerial photographs is easier than that of indoor and outdoor scenes where three-dimensional structures and arrangements of objects are crucial. (ARGOS system at Carnegie-Mellon University [66] tried to determine the view angle of aerial photographs of urban areas by using various two-dimensional projections of the three-dimensional configuration on the ground surface. The aerial photographs used in this system, however, are taken from very low altitude and hence can be considered as outdoor scenes rather than aerial photographs.)

But when we are going to analyze aerial photographs, we find several difficulties which are not encountered in other image analysis areas. Some of them are:

1. The picture is very large. In addition, since we usually have multiple pictures of a scene taken in different

spectral bands, the amount of picture data is very
large.

2. Because of the variety of photographic conditions
   under which aerial photographs are taken (season,
   weather, time, and altitude), the quality of a
   picture is apt to change. Therefore, it is almost
   impossible to make use of predetermined parameters
   for the processing of raw picture data.

3. The textural properties as well as the sizes and
   shapes of the objects vary quite widely, so that we
   have to incorporate sophisticated picture processing
   techniques to characterize these features.

4. The variety of objects in a scene which belong to
   quite different categories require diverse knowledge
   of the world.

5. There are so many different situations on the ground
   surface that it seems hardly possible to establish a
   model of the scene which represents all the possible
   mutual relationships among the objects.

The most serious of the above difficulties is that we are
unable to build a world model which represents the quite diverse
knowledge of the objects and all the possible spatial arrangements
among them. In the research on image understanding so far, the
interpretation of the scenes has been guided by models stored in
various forms: a constraint table [75], a set of probabilities [84],
a relational graph [3], and a semantic network [68]. Kanade [39]
pointed out that the first two representations of knowledge could
neither represent the hierarchy of the knowledge nor permit explicit
processing of the shape of object and object-dependent processing on
a part of the image. In order for the network representation of
knowledge to be substantially effective, objects in a scene should
have close contextual and semantic relations with each other. In
the case of aerial photographs, however, these constraints among
objects are restricted to be quite local, and some objects do not
have any constraints on others. For example, there are no contextual
and semantic relations between houses and crop fields, roads and
forest areas, etc. Thus the knowledge sources for each object tend
to be independent of each other. Therefore, the problems in the
knowledge-based analysis of aerial photographs are how to organize
such mutually independent knowledge sources and how to use them in a
flexible way as well as what knowledge sources are to be used.

1.5. Structural Analysis of Complex Aerial Photographs

In order to solve the above-mentioned problems in picture processing and knowledge representation, we have developed a system for

the structural analysis of complex aerial photographs based on "the production system" architecture. (A general discussion on the production system will be given in Section 2.3.) This system is an image understanding system which automatically performs knowledge-based analysis of complex aerial photographs in urban and suburban districts.

The major characteristics of our system can be summarized as follows:

1. The system has the ability to determine automatically the various parameters used in the processing of raw picture data. Thus, the system can give stable results despite the change of photographic conditions.

2. The system does not rely so much on the spectral properties of each point in a picture, but mainly uses spatial characteristics of regions to locate various objects. Hence, the system can give quantitative measurements related to the objects in a scene: area size, location, and number of objects.

3. Several experiments have been made to determine stable spectral characteristics of some materials (vegetation and water). The system incorporates this knowledge to characterize the properties of objects.

4. A method of extracting some three-dimensional information from an aerial photograph has been developed. That is, the system can estimate the locations of the objects with height by using shadow areas and the direction of the sun. This information is very valuable for discriminating three-dimensional objects such as houses, trees, and buildings from two-dimensional flat objects such as crop fields, grassland, and seas.

5. The system utilizes the knowledge about the locational constraints and the spatial arrangement rules to recognize context-sensitive objects. This ability has enabled it to recognize correctly cars on roads and regularly arranged houses.

6. As the size of a picture is very large, it would take a very long time to complete the analysis if the system applied sophisticated picture processing programs uniformly on the whole picture area. In order to solve this problem, the system incorporates a focusing mechanism, which a human being seems to use when he interprets a complex scene. First, the system estimates approximate areas where specific objects are highly probable. Then, it focuses its attention on those local areas and goes into the detailed analysis.

This not only saves very much processing time but also raises the reliability of object detection.

7. In the analysis of aerial photographs, the variety of objects in a scene and the irregularity of their spatial arrangements make it difficult and sometimes even impossible to fix the analysis process. Therefore, we have to develop a control mechanism where programs can be activated adaptively according to the structure of the picture under analysis. From this concern, we have adopted the production system as the software architecture of the system. The knowledge sources in our system are a group of "object-detection subsystems" which individually search for specific objects by communicating with each other via a common data base called "blackboard". The production system architecture is a very valuable tool for organizing diverse aspects of knowledge when only sets of mutually independent partial knowledge of the world are available. The modularity of the system makes it easy to modify or augment the knowledge sources stored in the system.

In the next chapter, we will show the overview of the system and specifications of the aerial photographs used in the experiments. The principle of the focusing mechanism and the control structure of the system will also be given more in detail. In Chapter 3, several picture processing algorithms used in the system will be described in detail together with some experimental results. Then, in Chapters 4 and 5, the algorithms for segmentation and extraction of characteristic regions used in the process of focusing will be presented. Chapter 6 is devoted to the detailed explanation of the algorithms for locating various objects. The control structure and the data organization in the system will be described in Chapter 7. In Chapter 8, we will evaluate the performance of the system by judging the results of the experiments made on several different aerial photographs.

2. OVERVIEW OF THE SYSTEM

2.1. Specification of the
Aerial Photographs Under Analysis

Before the development of remote sensing technology, only black and white (color) aerial photographs were available for surveying situations on the ground, and the analysis of the photographs was made by human specialists via visual inspection. This process, however, not only requires very specialized skills but also tends to be subjective. That is, the result of the analysis relies heavily on subjective judgements by an analyst and hence lacks objectivity. The development of various sensors for remote sensing has enabled us to measure multispectral properties of objects over a wide range of the spectrum, and techniques for pattern recognition and digital picture processing have facilitated the quantitative measurements of the situations.

Multispectral scanners and multiband cameras are the most popular sensors for recording the images of the ground surface. (Recently active sensors, such as side-looking airborne radar, have come into practical use for some special purposes.) For quantitative measurements of the situations in complex urban and suburban areas, camera data are preferable to scanner data because the former can provide the images in higher resolution and with less geometric distortion.

Urban and suburban districts show very complex land-use patterns where houses, buildings, roads, agriculture fields, forests, and so on are intermingled in an intricate fashion. As a result, we cannot identify each object individually in the images taken from high altitudes. Thus four-band multispectral camera data, taken from a low altitude by an airplane, are often used for a detailed examination of the districts. Figure 2.1[1] and Table 2.1, respectively, show one of the pictures and specifications of the aerial photographs used in this

---

[1] See color insert for Fig. 2.1.

study. This picture shows a typical landscape of a suburban area in Japan, which contains crops fields, forest areas, grassland, roads, cars, and many small houses. Hereafter we shall use this picture to demonstrate various algorithms used in the system.

Multispectral camera data consist of multiple gray-level pictures of the same landscape which are taken in different electromagnetic spectral bands. The photograph of each spectral band is sampled on a 256 × 256 grid and quantized to 256 gray levels. One pixel in a digital picture corresponds to 50 × 50 cm on the ground. As the area under analysis is very small, neither radiometric nor geometric corrections need to be carried out. After registration, four digital pictures of an aerial photograph in the BLUE, GREEN, RED, and INFRA-RED bands are passed to the analysis system.

2.2. Processing Sequence in the System

Figure 2.2 shows the block diagram of the system, where many processing modules are arranged around a common data base (blackboard). The analysis process in this system is divided into the following steps:

1. **Smoothing.** Four digital input pictures of an aerial photograph are smoothed by a sophisticated smoothing program name "edge-

Table 2.1 Specifications of the aerial photographs under analysis

DATE	Nov. 7, 1973		
CAMERA TYPE	4-band Multispectre Camera (I^{2}S model MK-1)		
ALTITUDE	1500 m		
SCALE	1 : 10000		
FILTER	B	W. #47A + IR BLOCK	
	G	W. #57A + IR BLOCK	
	R	W. #25 + IR BLOCK	
	IR	W. #88A	
LOCATION	Chiba Prefecture Japan		
DIGITIZATION	Drum Type Film Scanner (Mitsubishi Co.)		
	Spot Size : 50 µ		
	Sampling Pitch : 50 µ		

preserving smoothing" (the detailed algorithm of this smoothing will
be given in Section 3.1.). This smoothing not only removes noise in
homogeneous areas, but also reduces blur at edges, which facilitates
the subsequent segmentation process.

   2. **Segmentation.** After noise removal, the smoothed image of the
aerial photograph is segmented into many "elementary regions" according to the multispectral properties of each pixel. The segmentation
algorithm is based on a simple region-growing method, and does not
incorporate any knowledge of objects. These elementary regions,
where all the pixels have similar spectral characteristics, are considered as the basic units for the subsequent higher-level analysis.
Several basic properties of each elementary region are calculated
and stored in the blackboard.

   3. **Global survey of the whole scene.** Several kinds of regions
with characteristic properties, which we call "characteristic
regions", are extracted from the segmented picture. These characteristic regions are used to estimate approximate domains of objects.
The extraction of the characteristic regions is performed in parallel
by a set of "characteristic region extractors" (Fig. 2.2). In this
process no knowledge about specific objects is used. The only knowledge involved is the general knowledge about the aerial photograph
under analysis, such as photographic conditions and physical characteristics of the pictorial data: the direction of the sun, the size
of a pixel on the ground, spectral bands of the aerial photograph,
and so on.

   4. **Detailed analysis of focused local areas.** After extraction
of characteristic regions, a set of "object-detection subsystems"
perform the knowledge-based analysis to locate objects in a scene.
Each object-detection subsystem focuses its attention on specific
local areas by combining several characteristic regions (the detailed
discussion on the focusing mechanism will be given in Section 2.4.).
Then it checks the existence of specific objects by consulting the
knowledge stored in the subsystem. It is just this process where the
knowledge about the intrinsic properties and the environments of
specific objects is incorporated into the analysis.

   5. **Communication among object-detection subsystems.** All the
information about the properties of elementary and characteristic
regions and recognized objects is stored in the blackboard (Fig.
2.2). Each object-detection subsystem interfaces with it in a uniform way to check the conditions for activation and to write in the
results of analysis. All the subsystems communicate indirectly via
the blackboard. The system controls the overall flow of the analysis
by managing the information in the blackboard. It solves conflicts
among object-detection subsystems, and corrects errors in segmentation by backtracking to low-level processing.

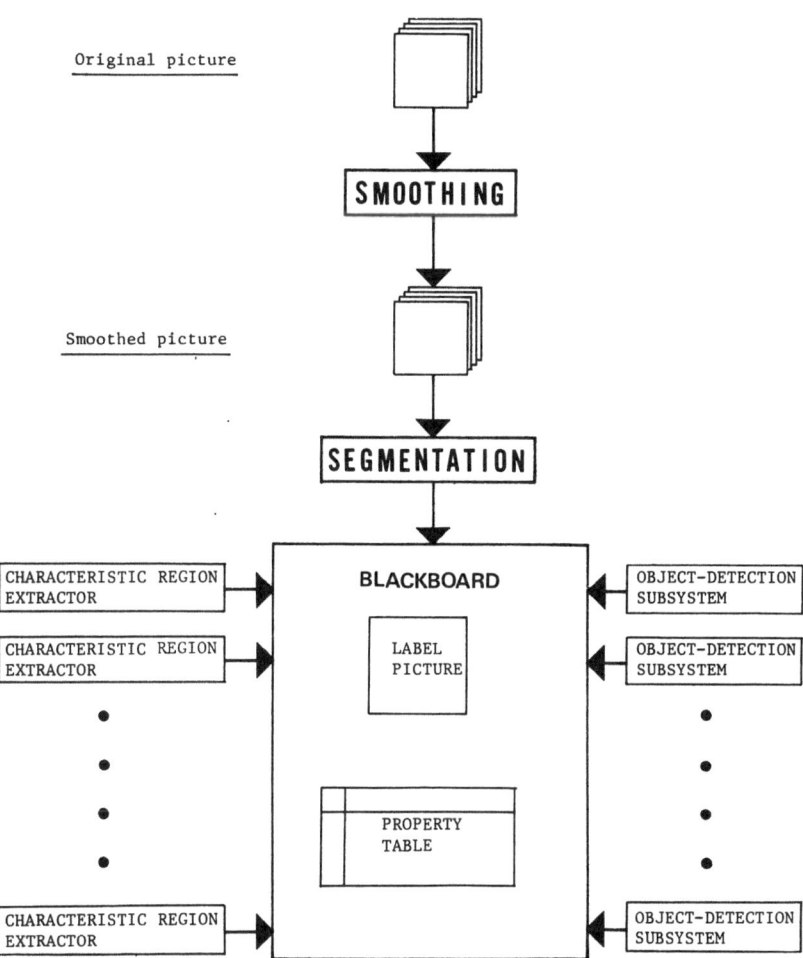

Fig. 2.2 Block diagram of the system

2.3. Production System as the Software Architecture

In this and the next sections we will describe in more detail
the software design of the system, i.e., the production system
architecture, and the idea behind the focusing mechanism.

2.3.1. Methodology in a Production System

Initially a production system was proposed as a general computation mechanism with the same ability as the Turing machine. It consists of three basic components: a set of rules, a data base and an
interpreter for the rules. Each rule consists of a pair of descriptions about the conditions for its activation and the actions on the
data base. The data base is the sole place in the system for storing
all the information to be recorded. The interpreter iterates the
following steps ("the recognize-act cycle") to proceed with the
computation.

1. Choose a set of rules ("the conflict set") whose
   conditions are satisfied by the current state of the
   data base.

2. Select one rule among the conflict set according to
   some criterion. (This process is usually called "the
   conflict resolution".)

3. Execute the actions described in the selected rule.

When the actions of the rule are executed, the contents of the data
base are modified, which results in triggering other rules and
restarting a recognize-act cycle.

For an example of the computation in a production system, suppose
that the data base contains a character string "ABC" and that the rule
set consists of the following rewriting rules.

Rule 1:   AB → BA

Rule 2:   BC → CB

Rule 3:   AC → CA ,

where the left-hand side of each rule denotes a condition to be
matched for its activation and the right-hand side of action to be
performed on the data base. At the first cycle of computation, the
conditions of Rule 1 and Rule 2 are satisfied by the content of the
data base. Then the interpreter selects one of them according to

some predetermined criterion. Here we suppose that the rules are ordered according to priority (Rule 1 has the highest priority), and that Rule 1 is chosen. As the result of activating Rule 1, the content of the data base is changed to "BAC", which results in activating Rule 3 at the second cycle of computation. At the end of this cycle the data base contains "BCA", and now Rule 2 is activated. Finally the contents of the data base are changed to "CBA".

The characteristics of a production system are summarized as follows [17]:

1. **Indirect channel of interaction**: No rule ever calls nor is it called directly from other rules. Interactions between rules are made indirectly via the data base.

2. **Modularity**: Each rule describes a primitive action for a task domain and is mutually independent of the others. As a result, a program becomes highly modular; one can change any rule without causing unexpected changes to other rules.

3. **Data-driven control**: Any rule can be activated at any time depending on the state of the data base. Nothing is prespecified about the manner in which rules will be employed in the computation. It is the information in the data base that determines the sequence of rule activations.

4. **Readability**: All the knowledge is explicitly described in a set of production rules without being concealed in the processing flow of a program. Therefore one can easily read the knowledge incorporated in the program merely by examining a set of rules.

Nowadays the production system finds various applications in "knowledge engineering", where various knowledge-based expert systems (e.g. DENDRAL for finding chemical structures [19], MYCIN for medical consultation [70], HEARSAY for speech understanding [42], etc.) are developed to perform the tasks which require human intelligence. In task domains such as perceptual psychology and clinical medicine, the knowledge of human specialists itself is not organized in a unified style. (Compare the situations in these fields with those in physics and mathematics.) All that we have is a collection of unrelated knowledge sources with less explicit mutual relationships. Consequently, it is very difficult or almost impossible to incorporate the knowledge in a computer program in a sufficiently well-ordered manner such as a decision tree, where the knowledge is highly organized in a hierarchical style. In these cases the production system offers a better framework for representing and structuring the diverse knowledge of human beings. Each rule in the production system represents a chunk of knowledge from a task domain, and is independent of the others.

Therefore it is quite easy to augument the performance of the system by adding new knowledge as another rule each time that a new insight of the world is acquired. This characteristics is quite valuable, especially in a task domain where one needs to develop a computer program that is evolutionary via repeated experiments, due to the lack of a coherent theory.

2.3.2. Production System in the Analysis of Aerial Photographs

As an aerial photograph contains a variety of objects such as crop fields, forest areas, grassland, rivers, roads, houses, and so on, the diverse knowledge about these objects should be incorporated to describe the structure on the ground surface. In addition, these objects, especially in urban and suburban areas, are intricately arranged without definite spatial relationships. Thus, the knowledge referring to some object tends to be independent of the knowledge of others. Moreover, an arrangement of the objects in a scene changes very widely from image to image. Therefore, it is almost impossible to build a cohesive world model which represents the diverse knowledge of objects and all possible situations on the ground surface. Taking these conditions into account, it seems to be natural to divide the system into a group of object-detection subsystems. Each of them is designed to locate specific objects using the knowledge of their intrinsic properties and the environments in which they are embedded. Thus, the diverse knowledge of the world is distributed in object-detection subsystems, each of which can be considered as an independent module of the system.

In such a distributed system configuration, the scheduling and the communication among modules become the most serious problems when designing the control structure of the system. That is, how can we determine the order in which a set of modules are activated, and how do these modules send messages to others and get the information about the environments in which they work? Some unified methodology has to be introduced to satisfy these questions.

Considering these requirements in the analysis of aerial photographs, we adopted the production system as the software architecture of the system. Each object-detection subsystem corresponds to a rule in the production system and represents the special knowledge required to locate a specific object in some particular environment. It becomes activated when the conditions of the rule are satisfied. The system need not schedule the subsystems since the data under analysis exactly determines the order of the activation of the subsystem.

All the information about picture data and recognized objects is stored in the common data base named "blackboard" [42]. Each object-detection subsystem interfaces with it in a uniform way, and checks the conditions for activation and writes in the result of the

analysis. It neither calls other subsystems nor passes messages to
them directly, because it does not notice the existence of the other
subsystems. The communication among object-detection subsystems is
made indirectly via the blackboard.

The modularity of the system makes it quite easy to modify,
delete, or add an object-detection subsystem without causing any
side-effects to the other subsystems. Therefore, when we obtain new
knowledge about the world, we can easily add it to the system and
increase its performance.

Although the original concept of a production system gives the
basic ideas of the knowledge representation and the control structure
of the system, we have to make several decisions on the detailed
design in order for the system to be suitable for the analysis of
aerial photographs: 1) The nature of processing performed by each
production rule, 2) The size of each production rule, 3) the goal-
oriented analysis, and 4) The mechanism of the conflict resolution.
(The detailed descriptions of the data structure in the blackboard
and the control mechanism of our system will be presented in Chapter
7.)

1. What kind of processing is performed by a production rule?
The original production system was designed as a general computation
scheme to process symbolic data, so that all processing performed in
the system can be decomposed into sets of simple symbol manipulations
such as replacement, deletion, insertion, and so on. But when we are
going to introduce a production system for image understanding, we
have to process picture data which is usually given in the form of
arrays of observed numerical values. Since the raw picture data is
not amenable to the intellectual processing, usually it is first
transformed into some forms of symbolic data by low-level picture
processing routines, and then analyzed by the knowledge-based high-
level processing routines. In order for the recognition process to
be completely symbol based, all features and relations among symbols
(elementary regions in our system) should be calculated and stored
in the data base in advance. As will be discussed in the next section, however, this takes a prohibitively long time and is quite
wasteful. Therefore, in order to save processing time, specialized
features required to recognize a specific object are to be calculated
only when they become necessary in the recognition process. (This is
a reason for the focusing mechanism of our system.) As a result, the
recognition process, which is done by a set of production rules, has
to perform picture processing, such as feature extractions, as well
as the symbolic data manipulations.

2. How much knowledge is to be represented in a production rule?
Since in the production system no control flow can ever be specified
explicitly, each production rule should be designed to perform a
meaningful unit of processing without expecting help from other rules.
What is important here is that each rule should represent a unit of

intelligent behavior in the task domain under consideration. Thus, the size of the knowledge source to be represented in a rule is very important to make the best use of the characteristics of the production system.

If we represent the whole knowledge about an object of some kind in a single production rule, the rule itself will become too big and complex to be managed, and the cooperative processing scheme in the production system will be greatly diminished. On the other hand, if a rule is very small and merely represents a method to calculate a specific feature, our knowledge of an object will be dispersed in several rules, and it will become very difficult to obtain an insight as to what knowledge is incorporated in the system.

Considering these conditions, we have decided to represent a specialized method to locate a specific object as one rule. Since there are several different ways to recognize a specific kind of object, the system has multiple rules for the recognition of a specific object. They work cooperatively to realize an efficient and reliable detection of the object. For example, our system contains four subsystems for the detection of houses, each of which represents the specialized knowledge for locating houses.

3. How can the system perform a goal-oriented analysis?
Researches on picture processing and image understanding so far have shown that simple picture processing routines cannot work well in complex natural scenes and that a goal-oriented processing is crucial for reliable analysis. In a production system, all the information is stored in the blackboard and each rule can read it at any time, so that we can easily implement a rule which performs a goal-oriented analysis. That is, once some object-detection subsystem locates an object, the properties of the object are stored in the blackboard. This enables other subsystems to utilize that information to detect "ambiguous" objects which were not recognized because of lack of reliability. A premature recognition of such ambiguous objects would have created a lot of "false objects". Consequently subsystems can perform the analysis quite reliably, being guided by a model of the object just detected from the scene. This increases both the efficiency and the reliability of object detection, and enables the system to make the best use of such heterarchical analysis to increase its performance. For optimum performance, initially all thresholds used for object discrimination are set very strictly to recognize only objects that are highly likely to be real. Then, the properties of the recognized objects are used to locate the rest of the objects.

4. How is the conflict between rules resolved? — Control strategy in a production system. In an ordinary production system the interpreter selects one rule from a set of rules ready to fire on the basis of some criterion (conflict resolution). The activation of the selected rule results in modifying the state of the data base. Then another set of rules become ready to fire. The computation in

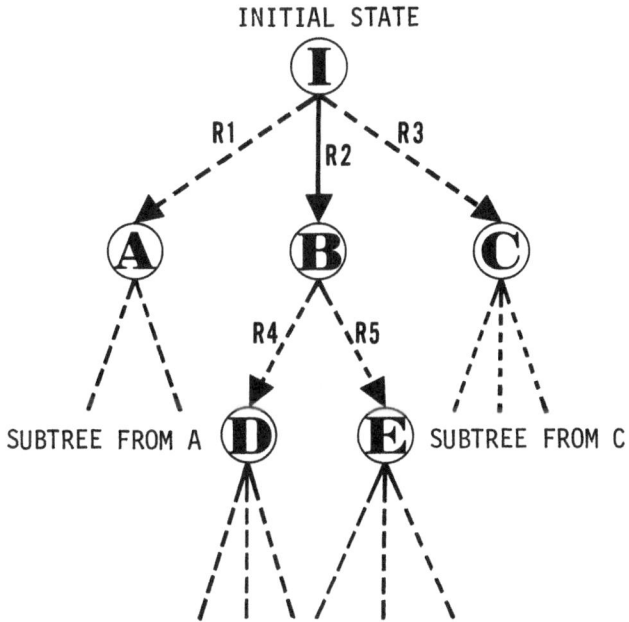

Fig. 2.3 Search tree expanded by a production system.

a production system is performed by the repetition of this "recognize-act cycle". Figure 2.3 shows a "search tree" expanded by this control strategy. There are three rules applicable to the initial state I of the data base. Suppose that the rule R2 is selected as the result of conflict resolution and that the state of the data base is changed to B. (The states A and C denote those to which the state I would be modified by the activation of the rules R1 and R3 respectively.) Then one of the rules applicable to the state B (R4 and R5) is selected and the state will be changed to D or E depending on which rule is selected. As is obvious from this search tree, subtrees from A and C are never expanded. Therefore this search strategy does not always give the optimal solution because the subtrees from A and C may contain a better solution than that obtained by expanding subtrees from B. Thus some sophisticated control mechanism has to be devised to obtain the optimal solution.

One way of solving this problem is to store all alternative states in the data base. That is, the data base stores a search tree

itself and the interpreter expands the most promising state in the
search tree. In the above example, even after applying the rule
R2, the rules R1 and R3 are ready to fire because the state I in the
data base still satisfies their conditions for activation. Then the
interpreter activates one rule among the rules R1, R3, R4, and R5
which seems to give the most promising solution. The HEARSAY II
system [42] incorporated this control structure, where the data base
(blackboard) represented alternative hypotheses in an integrated
manner. The problem here is the complexity of evaluating the order
of priority among rules ready to fire as well as that of information
organization in the data base.

The blackboard in our system also stores all alternative hypotheses asserted by a set of object-detection subsystems (production
rules) just like the HEARSAY II system. However, each object-
detection subsystem can be activated whenever its conditions are
satisfied regardless of the other subsystems. Therefore, our system
does not incorporate any criterion for scheduling the activation of
object-detection subsystems. This is because in aerial photographs
the analysis of objects of one kind can be performed rather independently of that of other kinds. For example, the recognition of houses
usually does not affect that of crop fields and forests. Therefore
the analysis by object-detection subsystems is performed in parallel
and the system need not make a complex scheduling among them. However, as each object-detection subsystem recognizes objects independently of the other subsystems, a region sometimes happens to be
recognized as different objects by multiple subsystems. For example,
it can happen that a region is recognized as both a crop field and a
grassland at the same time. Thus our system incorporates a mechanism
to resolve contradictions of this kind. The system always examines
the contents of the blackboard. If it finds a region which is recognized as multiple objects, the system evaluates the reliability of
the recognition of each object in some way or other, and it cancels
the recognition of those objects except for the most reliable one.
As a result, the interpretation of every region is uniquely determined
in the final result of the analysis. In what follows, we will call
this function of the system the "conflict resolution".

2.4. Focusing Mechanism of the System

As mentioned before, each object-detection subsystem is rather
complex and performs picture processing. Thus the recognition process would require a very long processing time if some mechanism to
reduce the computation time were not incorporated into the system.
In this section, we will discuss the focusing mechanism which has
been developed to realize an efficient analysis despite the complexity
of the processing in object-detection subsystems.

The size of a picture of an aerial photograph is very large and
a huge number of pictures are routinely to be analyzed. Therefore

the analysis of the picture should be done as efficiently as possible. If each very sophisticated and time-consuming object-
detection subsystem were uniformly applied to the whole picture
area, it would take a prohibitive time to complete the analysis.
Such a uniform analysis will require a long processing time to calculate all features of all regions in a picture, whilst most of
these features may often be useless because the features required
to recognize an object of one kind may be quite different from those
required for others. That is, it is very wasteful to apply a specialized feature extraction routine to such areas where an object
requiring that feature for its recognition is obviously absent. The
problem then is how to confine the complex analysis to very small
restricted areas where we are actually able to apply complicated
processing to locate the objects we want to obtain. In order to
solve this problem, we adopted the principle of a focusing mechanism
which a human being seems to use when he interprets a complex scene.

The perception process when a human is viewing a scene is very
complex. It has not been established how it functions, what he sees
in a picture, or how he understands the whole scene. But it seems to
be almost certain that when he sees a scene, he at first surveys it
globally to find prominent features which attract his interest. Then
he goes into a detailed examination of some local area to find objects
by using his knowledge of the world. The more intensively he focuses,
the more specialized knowledge he comes to use.

In the early stage of the focusing process we usually use neither
syntactic nor semantic knowledge of objects, because with the first
glance at a scene we only notice the existence of outstanding features
and we do not know *a priori* what is present in the scene. The result
of this process can be considered as a coarse description of the scene,
which guides the subsequent detailed examination.

When the analysis stage becomes detailed, the scope of the analysis is focused on some local area, and the specialized knowledge suitable for the properties of that area is introduced to search for
objects. Thus, the interpretation of the picture proceeds in a heter-
ogeneous manner; the analysis process of one local area is quite
different from those of others. For example, we check the textural
properties to classify trees in forest areas and count rectangular
regions in residential areas to obtain the number of houses, and so
on.

Kelly's [40] attempt of "planning" for the extraction of edges
may be regarded as one of the first trials of the focusing of attention in picture processing. His planning consisted of three steps:

1. Make a new small picture by reducing an original picture.
   Each point in the reduced picture is given the average
   gray value of the points in a nonoverlapping N × N
   square in the original picture (the resolution is
   reduced by a factor of N).

2. Locate edges in the new picture.

3. Use the edges found in the reduced picture as a plan
   for finding edges in the original picture.

When the edges in the small picture are mapped back onto the original
picture, they are regarded as reference areas where detailed algorithms are applied to extract detailed and exact edges in the original
picture. (One point in the reduced picture corresponds to an $N \times N$
square in the original picture.) As the reduced picture is a
smoothed version of the original picture, the edge detection in that
smoothed picture becomes easier and more reliable. Moreover, as the
edge detection in the original large picture is now made only in
restricted local areas, the planning saves very much processing time.

Pyramid data structures (or processing cones) developed by
Tanimoto and Pavlidis [74] and others [27,79] may be considered as
the extension of Kelly's idea. The shrinking operation by a factor
of two is repeatedly applied to the original picture, resulting in
a hierarchy of averaged smaller pictures aligned like a pyramid
(Fig. 2.4). The edge detection in the pyramid is made by a "top-
down" analysis; at first, the edges in the smallest picture are
detected, and then the edges in the next larger picture are located
using those in the previous picture as the guide-areas, and so on.
Many researches using the pyramid data structure have shown its
effectiveness in picture processing and computer vision [61,65].

Although the pyramid data structure is useful when the picture
under analysis contains a few objects in a large homogeneous background, it has several serious drawbacks in the case of complex
scenes such as aerial photographs:

1. As the sizes and the textural properties of objects on
   the ground surface vary quite widely, it becomes very
   difficult to establish an optimum size of shrinking to
   detect clues to objects. In most of the cases, a
   reduced picture in the pyramid suitable for objects of
   one kind is not useful for detecting objects of other
   types. Therefore, complex adjustments among the pictures
   in the pyramid will be required to detect a variety of
   objects successfully.

2. An aerial photograph of urban and suburban areas contains objects almost everywhere. Thus, such blurring
   process often gives rise to false regions or sometimes
   smooths out small objects, which results in mis-
   leading the later stage of the analysis. For example,
   some closely located objects are often merged into one
   region in a picture at a higher level of the pyramid.
   In this case the resulting false region shows properties
   quite different from those of the original regions.
   Since we do not know in advance what objects are present

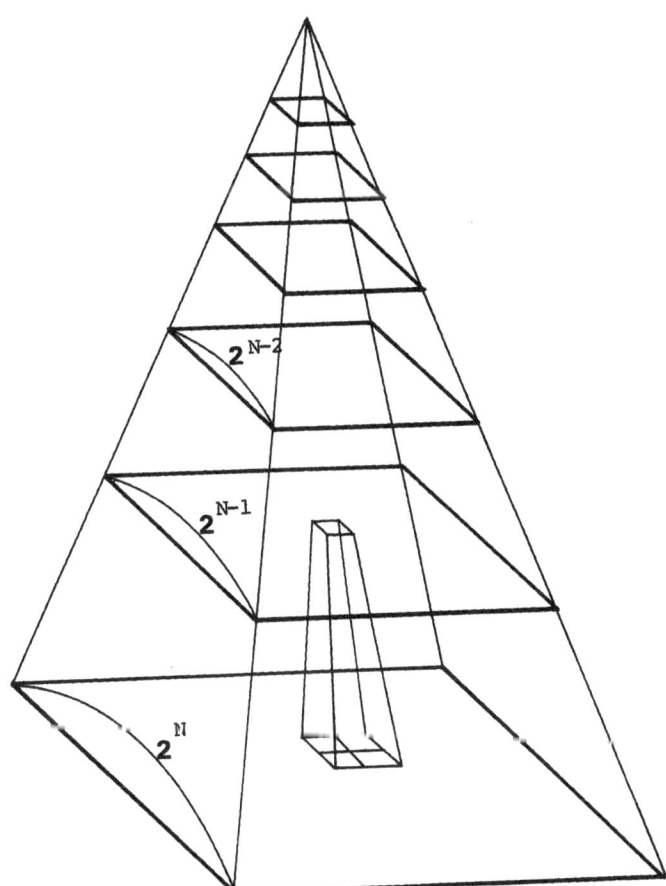

Fig. 2.4 Pyramid data structure

in a scene and where they are located, we have no way
of predicting at which level of the pyramid objects
get merged.

The principle of the focusing mechanism in our system is as
follows. In the pyramid data structure the raw picture data are
hierarchically organized along the resolution and the focusing
process proceeds to the detailed examination by increasing the resolution step by step. It utilizes only one feature (*i.e.* resolution)
to estimate approximate areas of objects. But there are many other
features useful for confining spatial domains of objects. For
example, long thin regions are very useful clues to locate elongated
objects such as roads, rivers, and railroads, and very large uniform
regions can be considered as large flat objects such as crop fields
and water surface. Therefore, as opposed to the hierarchical organization of picture data in the pyramid, our system organizes the
picture data in parallel according to several different characteristics such as size, shape, color, and texture.

Figure 2.5 shows the process of the focusing of attention in our
system. After segmentation several kinds of regions with prominent
features (characteristic regions) are extracted. These features are
distinguishing features of the objects to be detected. This stage
corresponds to a global survey of a scene and its result guides the
subsequent detailed analysis. Next, each object-detection subsystem
estimates approximate areas of the objects by combining several different characteristic regions. For example, a common area between
a heavily textured area and a green-colored area can be considered
as almost exactly corresponding to a forest area. Thus, each subsystem applies logical operations (such as "and", "or", and "negation") among various characteristic regions to extract spatial domains
of the objects. Then, it analyzes extracted areas in detail to locate
the objects consulting the specialized knowledge stored in the subsystem. The time-consuming processing by object-detection subsystems
is applied only in small local areas where specific objects are highly
probable, so that object detection becomes very reliable and the total
processing time can be very much reduced compared to a uniform overall
processing. This reduction in processing time is very great especially for a large picture such as an aerial photograph.

In this focusing process, in order to extract characteristic
regions, we have to analyze the whole picture area several times
But, since this organization process is in substance a parallel
processing, if needed we can implement a parallel processor to perform it at high speed. The processor consists of a set of processing
units sharing common memory, each of which extracts a specific kind
of characteristic region. Once various kinds of characteristic
regions are extracted, we can estimate spatial domain of various
objects via diverse combinations of characteristic regions. This
focusing mechanism in our system will be very useful for complex

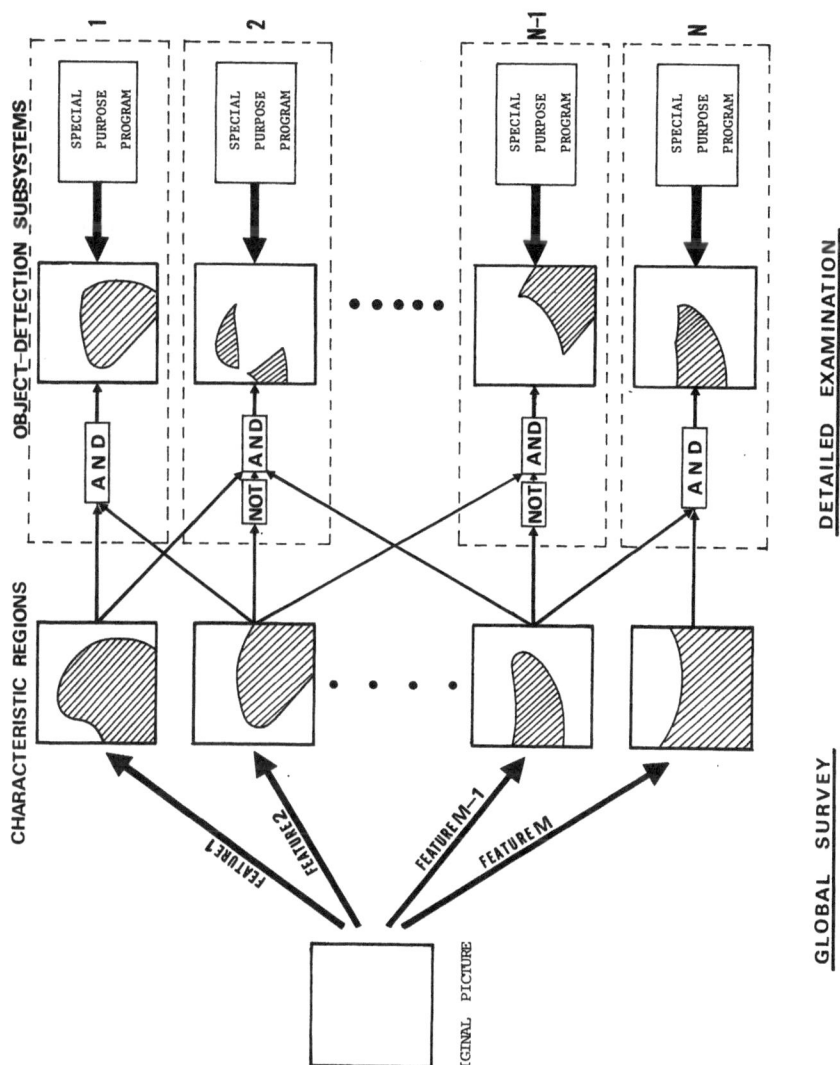

Fig. 2.5 Focusing mechanism of the system

scenes where a variety of objects with different properties are
present almost everywhere.

The idea of extracting characteristic regions has a feature
in common with the idea of "general purpose models" (GPM's) by Zucker,
Rosenfeld and Davis [85]. They stressed in their paper the importance of the knowledge-free feature extraction and data organization
as follows:

> "GPM's are models which are applicable even when we
> have little or no *a priori* knowledge about the class
> of scenes that is to be analyzed. They include models
> for general classes of local features that occur in
> many different types of scenes, as well as models that
> describe how such features can be grouped into aggre-
> gates. These aggregates may in fact not correspond
> to objects but they can serve as useful first guesses
> to guide later steps in the analysis .... The early
> stages of visual information processing do not depend
> on one's knowledge or expectations ("perceptual set")
> about the particular situation .... Thus, some of the
> operations performed by the human visual system can be
> thought of as corresponding to the use of GPM's at the
> early stages of scene analysis."

Marr's [46] idea of "primal sketch" may be regarded as being
along the same lines. All these ideas (including ours) can be summarized as follows. In image understanding, much work can be done
to organize the raw picture data before incorporating special task-
dependent knowledge. This organization process makes it easy to
isolate objects and directs later stages of the analysis. The more
complex a scene is, the more important this organization becomes.

The characteristic regions are extracted by a set of "characteristic region extractors" (Fig. 2.2). These routines describe the
coarse structure of the scene without using any knowledge about particular objects. They merely utilize physical characteristics of
the picture data under analysis (photographic conditions), general
models for local features (edges, lines, etc.), and various properties of regions (size, shape, etc.). In this sense, the characteristic region extractors can be said to represent the general knowledge required for structuring picture data, while object-detection
subsystems represent the specialized knowledge to locate objects in
restricted circumstances. Thus, the knowledge in our system is
organized on two different levels: the general and the specialized.

Each characteristic region extractor in our system examines
elementary regions in the blackboard which are generated at the segmentation process, and extracts a specific kind of characteristic
region independently of the others. Then it writes the result of
the analysis into the blackboard (Fig. 2.2). Therefore, if necessary,

we can change any routine without considering any side-effects on the others. In this sense, these routines might be regarded as the rules in the production system, though they are activated only once after segmentation regardless of the contents of the blackboard (*cf*. ordinary production rules can be activated at any time depending on the contents of the blackboard). Accordingly, when we obtain some new knowledge useful for the organization of raw picture data, we can easily incorporate it into the system to raise its reliability in estimating the spatial domains of objects.

3. SOME BASIC TECHNIQUES IN PICTURE
PROCESSING AND FEATURE EXTRACTION

Image understanding systems for the analysis of complex scenes should have highly developed abilities in picture processing and feature extraction to process digital images and measure the various properties of regions and lines. Especially in the analysis of aerial photographs, we have to calculate many different features to characterize a variety of objects: boundary smoothness for crop fields, elongatedness for roads, rivers and railroads, squareness for houses and buildings, textural properties for grasslands and forest areas, etc. The quality of these measurements, obtained by picture processing and feature extraction, has crucial effects on the higher-level recognition process. Thus, the early stages of feature extraction must be as accurate as possible to obtain good final results.

In this chapter we shall describe three sophisticated algorithms for picture processing which have been newly devised while developing our system: a smoothing method named "edge-preserving smoothing", a measure of elongatedness of a region, and a structural description method of regularly arranged patterns. These algorithms are used in our system of aerial photograph analysis for preprocessing raw picture data, feature extraction, and object recognition, respectively. How these algorithms are used in the system will be described in appropriate chapters. The reader who is interested in the major story of aerial photograph analysis alone may skip this chapter, and return here when the descriptions of the analysis processes using these algorithms appear.

3.1. Edge-Preserving Smoothing

There have been many papers on the subject of smoothing a digital image. [37,58,64,71,76] A basic difficulty of smoothing is that, if applied without care, it tends to blur any sharp edges which happen to be present. Edge-preserving smoothing is a sophisticated smoothing

algorithm which attempts to resolve the conflict between noise eli-
mination and edge degradation. It looks for the most homogeneous
neighborhood around each point in a picture, and then gives each
point the average gray level of the selected neighborhood. It removes noise in flat regions without blurring sharp edges or destroying the details of the region boundaries. All these characteristics
of edge-preserving smoothing will be quite valuable for segmentation
and edge detection.

3.1.1. Blurring Effect of Smoothing

The most straightforward way to reduce noise in a gray-valued
digital picture is to take local averages: each point in a picture
is given the average gray level of some neighborhood of that point
(N x N square area centered at each point is often used as the
neighborhood). This method, however, tends to blur sharp edges
between different regions; as the location of the neighbor
fixed, we have no way of keeping the neighborhood from including
edges, which results in blurring the edges. In order to avoid this
side-effect of local averaging, the averaging has to be done only
in a neighborhood which does not contain any sharp edges.

Recently Tomita and Tsuji [76] have proposed a smoothing method
which gives the point (X, Y) the average gray level of the most homogeneous neighborhood among the five square neighborhoods shown in
Fig. 3.1. However, their method does not yield a good result if
applied to a complex-shaped region, because this method uses the
square areas as the neighborhoods around (X, Y). For example, a
wedge-shaped portion of a region is apt to be merged into the surrounding regions, or sometimes it becomes an independent region with
some false gray level. Furthermore, an $N \times N$ region, whose size is
the same as that of square neighbourhoods used for the smoothing
cannot survive after several iterations of the smoothing operation.

Let us consider a simple example. Suppose that we apply
Tomita's smoothing method to the 3 × 3 region represented in Fig. 3.2
by using five 3 x 3 square neighborhoods. At the points A, C, E,
G, and I there exist square neighborhoods which are completely included in the 3 × 3 region, while at B, D, F, and H all five 3 × 3
square neighbourhoods lie over the boundaries, and contain parts of
both the region and the surrounding region. Therefore the new gray
levels given to these points become smaller (larger) than the original value of the region. If this process is iterated several
times, the gray levels of the points in the 3 × 3 region approach
that of the surrounding region, and finally this region is smoothed
out.

In order to avoid this effect it is necessary to determine a
new shape for the neighborhood in which the local average is taken
and also a new rule for iterating the algorithm.

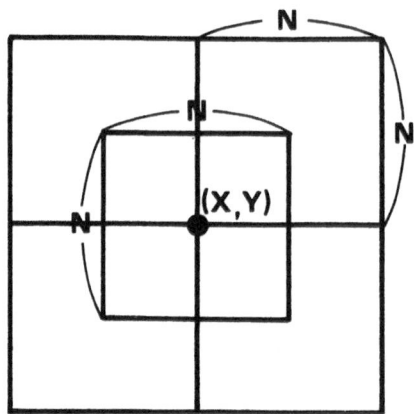

Fig. 3.1 Five square neighborhood areas around a point (X, Y)

● Represent pixel centers

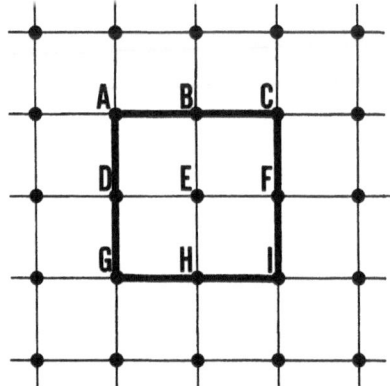

Fig. 3.2 A 3 × 3 region; the region consists of nine pixels
         (from A to I) and is surrounded with a background
         region (unmarked pixels)

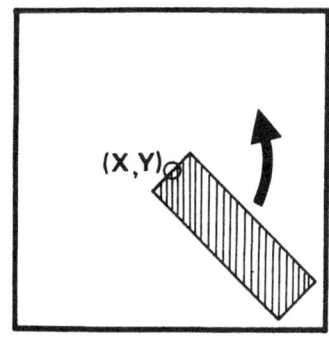

Fig. 3.3 Mask selection for
          edge-preserving smooth-
          ing; rotate an elongated
          bar mask around (X, Y)
          and detect the position
          of the mask where its
          gray-level variance is
          minimum

3.1.2. Edge-Preserving Smoothing

The procedure of our edge-preserving smoothing is as follows:

Step 1: Rotate an elongated bar mask around a point
         (X, Y) (Fig. 3.3).

Step 2: Detect the position of the mask where its gray-
         level variance is minimum.

Step 3: Give the average gray-level of the mask at the
         selected position to the point (X, Y).

Step 4: Apply the Steps 1 - 3 to all points in a picture.

Step 5: Iterate the above process until the gray-levels
         of almost all points in a picture do not change.

In order to remove the noise without blurring sharp edges,
averaging must not be applied to an area which contains an edge
because it blurs the edge. Thus the most homogeneous neighborhood
is to be found around a point to be smoothed. If an area contains
a sharp edge, the variance of the gray-level in that area becomes
large. Therefore we can use the variance as a measure of nonhomo-
geneity of an area.

Suppose that a picture has two regions R1 and R2 whose gray-
level means and variances are $(0, \sigma_1^2)$ and (m, σ_2^2) respectively.

Let a point (X, Y) belong to R1. If (X, Y) is located at the central part of R1, its gray-level approaches to the average gray-
level of R1 (in this case 0) after several iterations of smoothing.
On the other hand, if (X, Y) is near the boundary, there exist two
kinds of neighborhoods, one of which is completely included in R1

while the other includes parts of both R1 and R2. The variance of the former is about σ_1^2. The variance σ^2 of the latter can be calculated as follows. Let N_1 and N_2 denote the numbers of points belonging to R1 and R2, respectively, which are contained in this neighborhood. Then the variance σ^2 is about

$$\sigma^2 \simeq \frac{1}{N} \left[\sum_{i=1}^{N_1} \left(x_i^1 - \frac{N_2}{N} m \right) + \sum_{j=1}^{N_2} \left(x_j^2 - \frac{N_2}{N} m \right)^2 \right],$$

where $N = N_1 + N_2$, x_i^1 and x_j^2 denote the gray levels of the points belonging to R1 and R2, respectively, and $\simeq$ means nearly equal. Expanding the right-hand side,

$$\sigma^2 \simeq \frac{1}{N} \left[N_1 \sigma_1^2 + N_2 \sigma_2^2 + \frac{N_1 N_2}{N} m^2 \right],$$

where we have used

$$\sum_{i=1}^{N_1} \left(x_i^1 \right) \simeq N_1 \sigma_1^2, \quad \sum_{i=1}^{N_1} x_i^1 \simeq 0, \quad \sum_{j=1}^{N_2} \left(x_j^2 - m \right)^2 \simeq N_2 \sigma_2^2,$$

and

$$\sum_{j=1}^{N_2} x_j^2 \simeq N_2 m.$$

If $\sigma^2 < \sigma_1^2$, that is, $\sigma_2^2 + \frac{N_1}{N} m^2 < \sigma_1^2$, the neighborhood containing both parts of R1 and R2 is selected for averaging. The boundary between R1 and R2 is then blurred by the averaging operation over this neighborhood. In most pictures, however, it is reasonable to assume $\sigma_1^2 \simeq \sigma_2^2$, and then $\sigma^2 > \sigma_1^2$, so that the correct neighborhood is selected. Even if R1 is very noisy and σ_1^2 is large, σ^2 is sufficiently larger than σ_1^2, provided that the difference of the average gray values of the two regions, *i.e.*, m, is large.

3.1.3. Actual Implementation for Discrete Picture Data

The nine masks in Fig. 3.4 are the discrete realizations of
the bar masks of the smallest size for the edge-preserving smoothing of a digital picture. Using pentagonal and hexagonal corners
at the point (X, Y), we can avoid the degradation of sharp edges,
and can find the homogeneous neighborhoods (i.e., neighborhoods
with no sharp edges in them) even when the point (X, Y) is located
at an acute angle of a complex-shaped region. Thus we can smooth
a region without blurring sharp edges or destroying the shape of
the boundary.

The 3 × 3 square mask in Fig. 3.4 is used to smooth a small
region. Suppose that we apply this smoothing to the 3 × 3 square
region in Fig. 3.2   At the points A, B, C, D, F, G, H, and I, there
exists a pentagonal or hexagonal mask which is completely included
in that region, while at E all these eight masks contain both points
from the region and the surrounding area. If we add a 3 x 3 mask as
a neighborhood of the point (X, Y), then we can smooth even a 3 x 3
region without destroying the shape. The variances of these nine
masks are compared with each other, and the average gray level of
the mask with the least variance is given to the point (X, Y).

Generally, in order to significantly reduce the amplitude of
the noise fluctuations, the neighborhood for averaging should be
large. Smoothing by averaging over a large neighborhood, however,
smooths out small regions and destroys the details of the boundaries. When we use small neighborhoods (seven points for pentagonal
and hexagonal masks or nine for a square mask) around each point in
a picture, in order to preserve the details of boundaries, we cannot
reduce the noise significantly. However, the iteration of this
smoothing operation can produce the same or much better effect than
a single averaging over a larger neighborhood. Thus we apply this
program repeatedly until the gray levels of almost all points in a
picture do not change.

Figure 3.5 shows the results of the edge-preserving smoothing
of a simple pattern, which has been artificially generated on a
100 × 100 grid and quantized to 256 gray levels. The original picture has three different regions with the average gray levels 48,
108, and 168, respectively. Gaussian noise has been added to this
pattern with the standard deviations of (a) 10, (b) 20, (c) 30,
(d) 40, and (e) 50, respectively. The noisy pictures on the top
row of Fig. 3.5 have been made by this operation. The pictures on
the middle row show the results of the edge-preserving smoothing
after ten iterations. The pictures on the bottom row show the edges
of the smoothed pictures as the result of an ordinary differentiation and thresholding. The edges and the angles of each region are
preserved almost completely although the picture is very noisy.
The effectiveness of our edge-preserving smoothing at a sharp angle
can be seen by comparing Fig. 3.5 (a)—(e) with Fig. 3.6 (a)—(e),

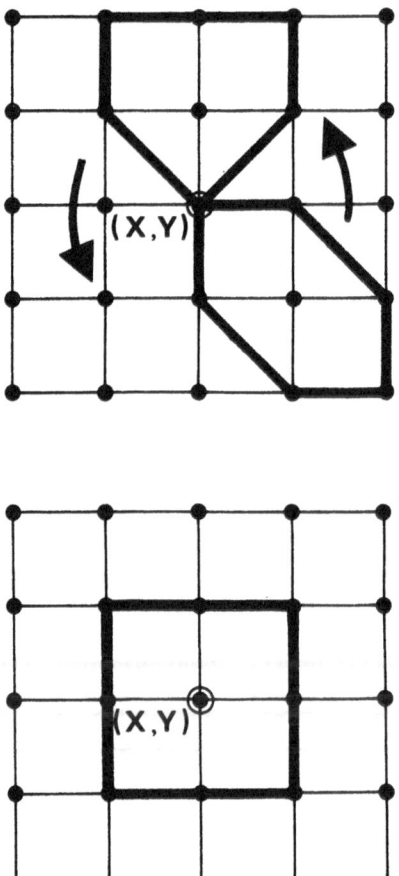

Fig. 3.4 Discrete realizations of the bar masks; four pentagonal
          and four hexagonal masks have sharp corners at the point
          (X, Y). A 3 × 3 square mask is used to smooth a small
          region

Fig. 3.5

Result of the edge-preserving
smoothing of a simple artificial
pattern which has three different
regions with the average gray
levels 48, 108, and 168, respectively. The pictures in the top
row are corrupted by Gaussian
noise with standard deviations
10, 20, 30, 40 and 50, respective
ly ((a) to (e)). The pictures in
the middle row are the results of
the edge-preserving smoothing
after ten iterations. The pictures in the bottom row are the
results of ordinary differentiation and thresholding of the
smoothed patterns. Isolated
noise is clearly removed and the
boundaries of regions are almost
completely preserved

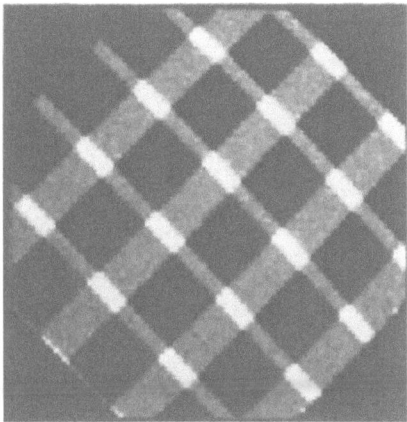

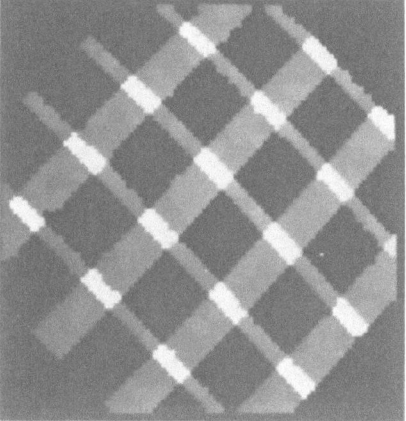

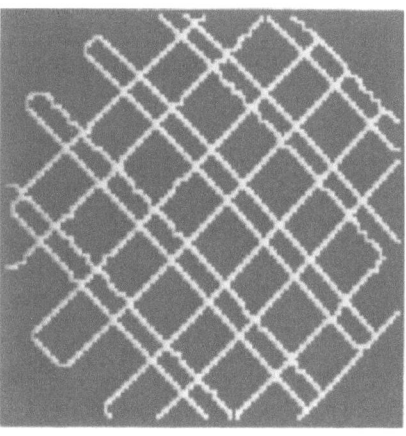

(a)

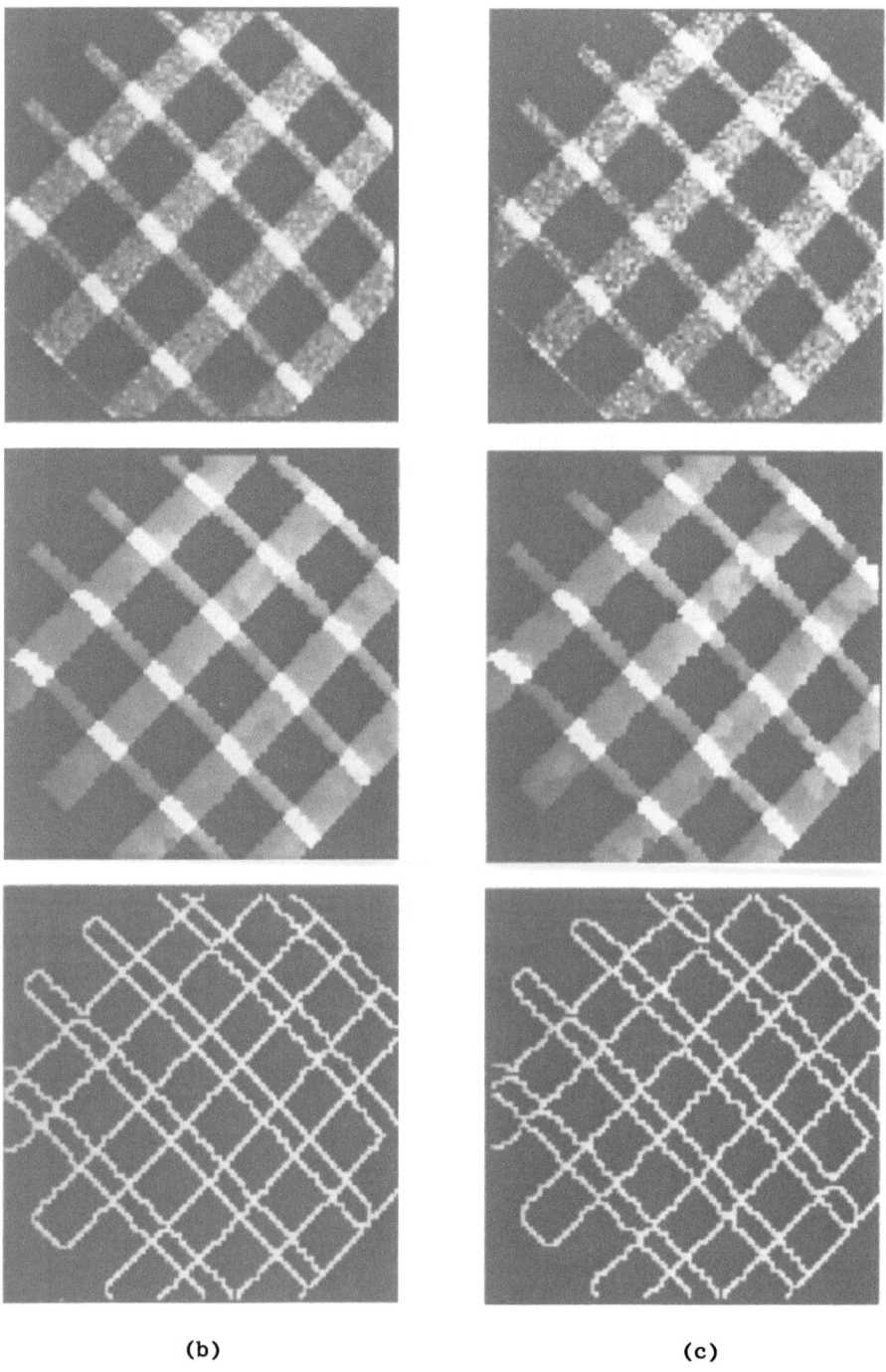

(b)   (c)

Fig. 3.5 contd.

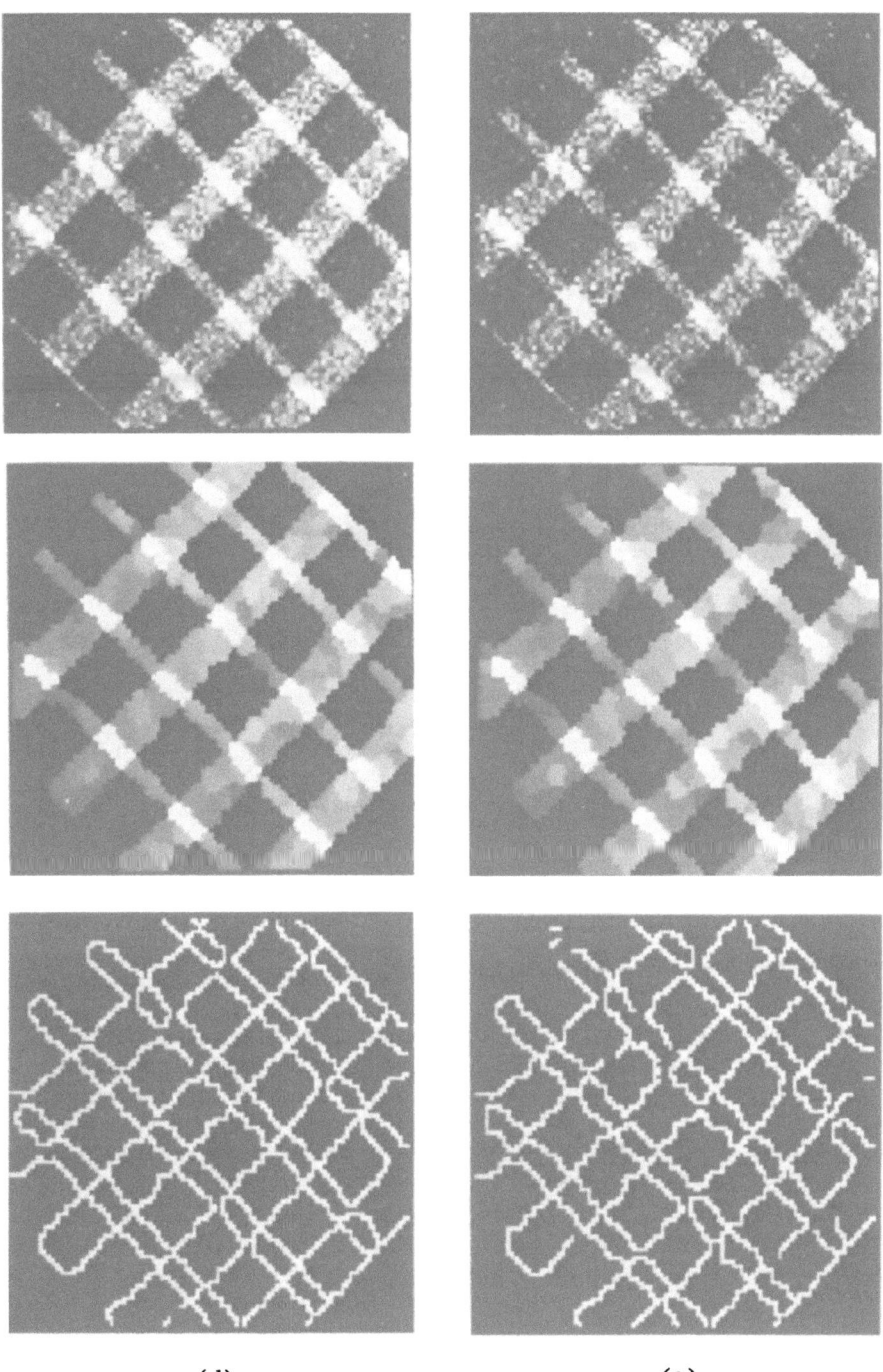

(d)   (e)

Fig. 3.5 contd.

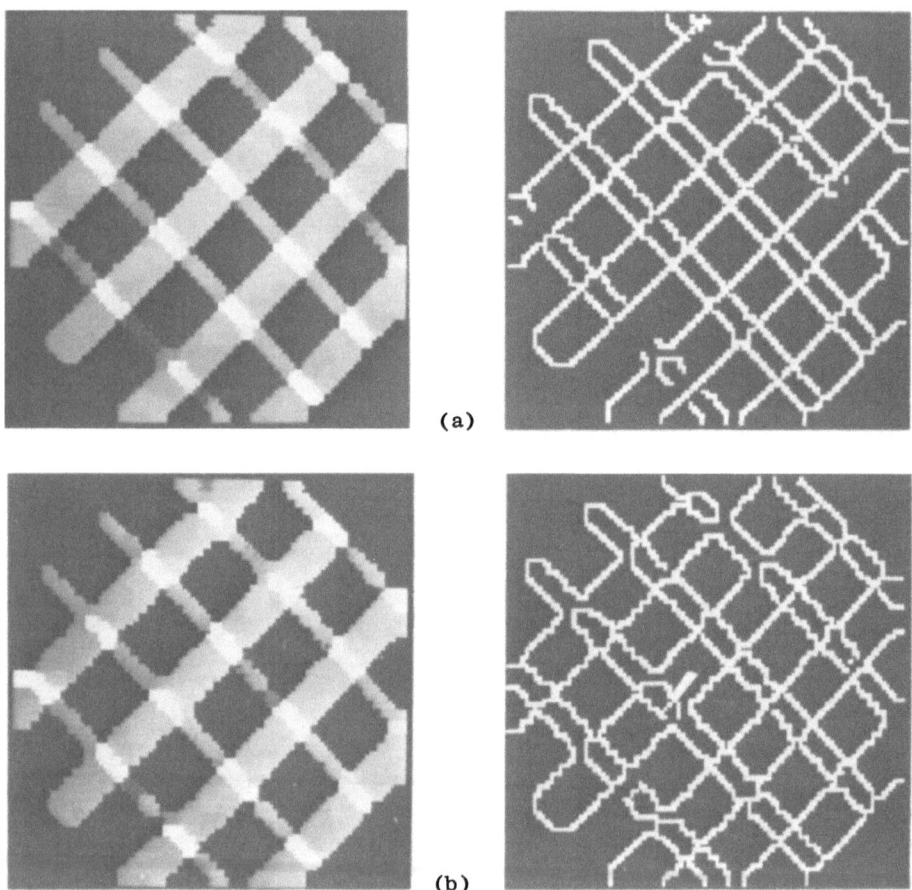

Fig. 3.6 (a) — (e) Results of smoothing the noisy artificial
          patterns of Fig. 3.5(a) to (e) by the method proposed
          by Tomita and Tsuji (upper row). Results of the same
          differentiation and thresholding as in Fig. 3.5 (bottom
          row). By this smoothing, sharp corners of regions are
          destroyed and several regions are smoothed out

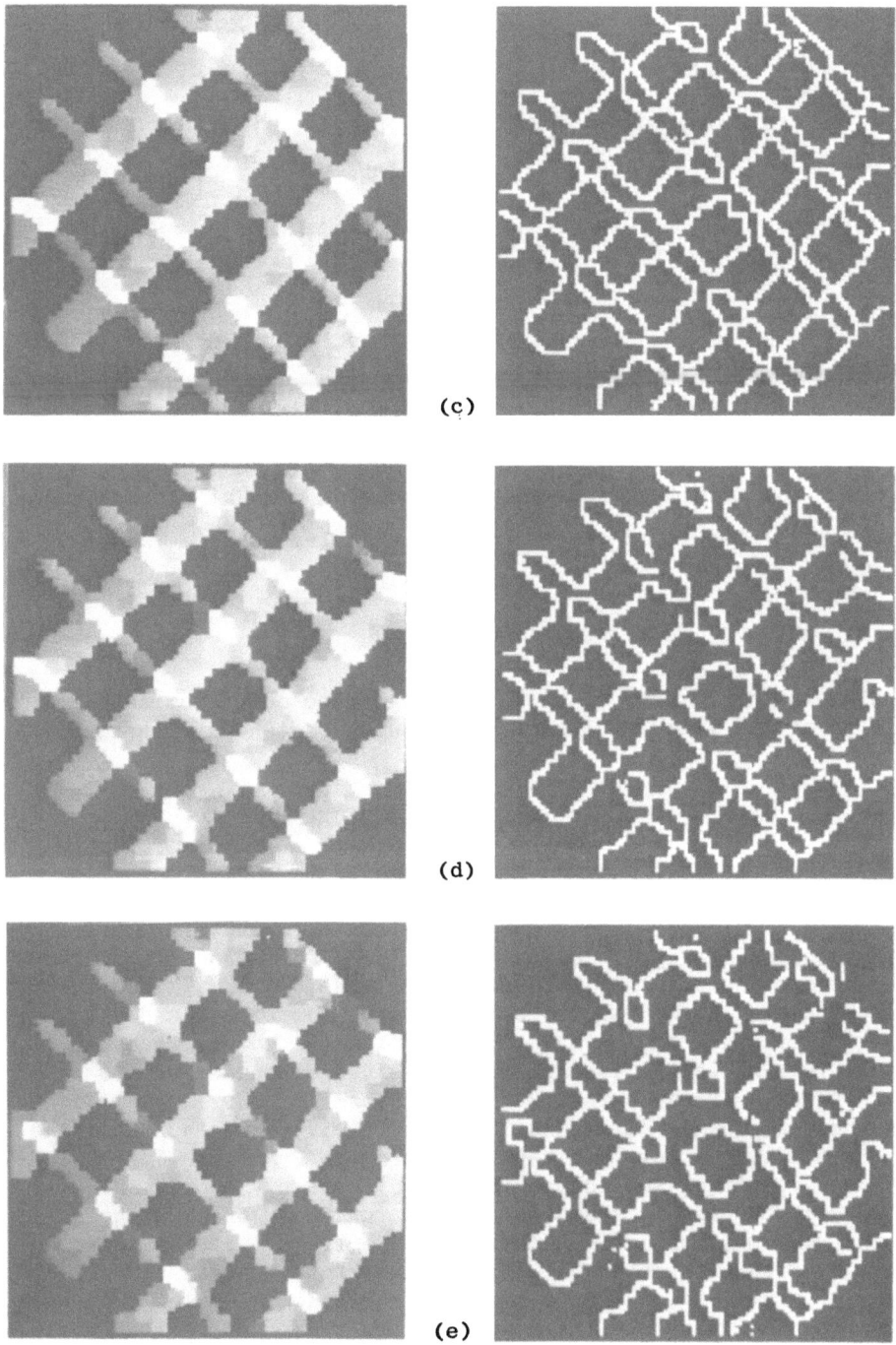

Fig. 3.6 contd.

GRAY LEVEL   0 , 0 , 0 , 0 , 1 , $\overset{a}{3}$, $\overset{b}{5}$, 8 , 8 , 8 , · · ·

(a)   MEAN   0 , 0 , 0 , 1/3 , 4/3 , 3 , 16/3 , 7 , 8 , 8 , · · ·

VARIANCE   0 , 0 , 0 , 4/9 , 14/3 , 8 , 38/3 , 6 , 0 , 0 , · · ·
(x 3)

(b)   0 , 0 , 0 , 0 , 0 , 1 , 7 , 8 , 8 , 8 , · · ·

(c)   0 , 0 , 0 , 0 , 0 , 0 , 8 , 8 , 8 , 8 , · · ·

Fig. 3.7 Sharpening of a blurred one-dimensional edge

(a) A blurred one-dimensional edge

(b) Result of smoothing the one-dimensional blurred edge

(c) Result of smoothing (b); the blurred edge is sharpened

which shows the results of the smoothing proposed by Tomita and
Tsuji using the five 3 x 3 square neighborhoods shown in Fig. 3.1.

3.1.4. Sharpening of Blurred Edges

This smoothing program not only removes the noise but also
sharpens blurred edges. This effect can be well understood by considering a simple one-dimensional example. Figure 3.7(a) shows a
one-dimensional digital blurred edge. The numbers under each gray
level denote the mean and the variance of the 3 × 1 mask centered
at each point. If we select the minimum-variance mask and give the
average gray level of the mask selected for the point, we obtain
Fig. 3.7(b), where the average gray levels are rounded off to in-
teger values. At the points "a" and "b" on the blurred edge we
have the new gray levels 1 and 7, respectively, which approach the
gray levels of the left and right flat regions. Figure 3.7(c)
shows the result of smoothing Fig. 3.7(b), where the original blurred edge is completely sharpened. This function works in the same
way on two-dimensional digital pictures. Figure 3.8 (a) and (b)
show a blurred artificial pattern and its cross section along the
diagonal line. Figure 3.8 (c) and (d) show the result of the edge-
preserving smoothing and the corresponding cross section. Blurred
edges are clearly sharpened.

3.1.5. Convergence

The fluctuations of the gray levels are gradually reduced by several iterations of the edge-preserving smoothing. Once a point has a neighborhood of constant gray level, its gray level is never changed by the smoothing because the variance in this neighborhood is always zero. Therefore the number of points whose gray levels are changed by the smoothing will gradually decrease to zero. (Ordinary local averaging also converges after many iterations. But this is meaningless because all points in a picture will have the same gray level.) Figure 3.9 shows a typical example of this phenomenon for the artificial pattern of Fig. 3.5(c), where the horizontal axis shows the number of iterations and the vertical axis the number of points whose gray levels are changed. Even though this curve does not decrease exactly to zero, the gray-level changes caused by an additional smoothing operation are very small, *i.e.*, at most one or two gray levels, and the process can be regarded as converged. In this simple artificial pattern, a few iterations are sufficient for practical use. The number of iterations needed for convergence depends on the amplitude of the noise fluctuations and the shapes of the regions in a picture.

3.1.6. Conclusion

We have shown a new smoothing algorithm named "edge-preserving smoothing". This smoothing removes noise without blurring sharp edges, and still has the ability of sharpening blurred edges. These excellent characteristics of the edge-preserving smoothing may better be appreciated by comparing its performance to other smoothing methods.

Recently the median filtering [37,58] has been proposed as a new smoothing method, which gives each point in a picture the median, instead of the average, of the gray levels in its neighborhood. Thus, it is said that median filtering can give better results than local averaging in that it causes less blur at sharp edges. We applied this median filtering to the artificial pattern with moderate noise. Figure 3.10 (a) and (b) show the original noisy pattern and its cross section along a horizontal line, respectively. Figure 3.10 (c) and (d) show the result of median filtering, and Fig. 3.10 (e) and (f) the result of edge-preserving smoothing. It is clear from these results that our smoothing is superior to the median filtering. In addition, one can easily see that the median filtering cannot sharpen blurred edges like the edge-preserving smoothing when applied to the blurred picture in Fig. 3.8.

The above-mentioned characteristics of the edge-preserving smoothing will be quite valuable for extracting homogeneous regions

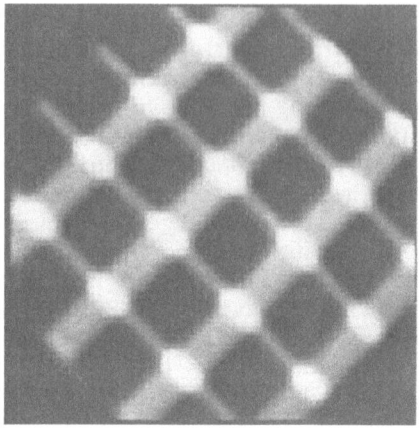

(a)

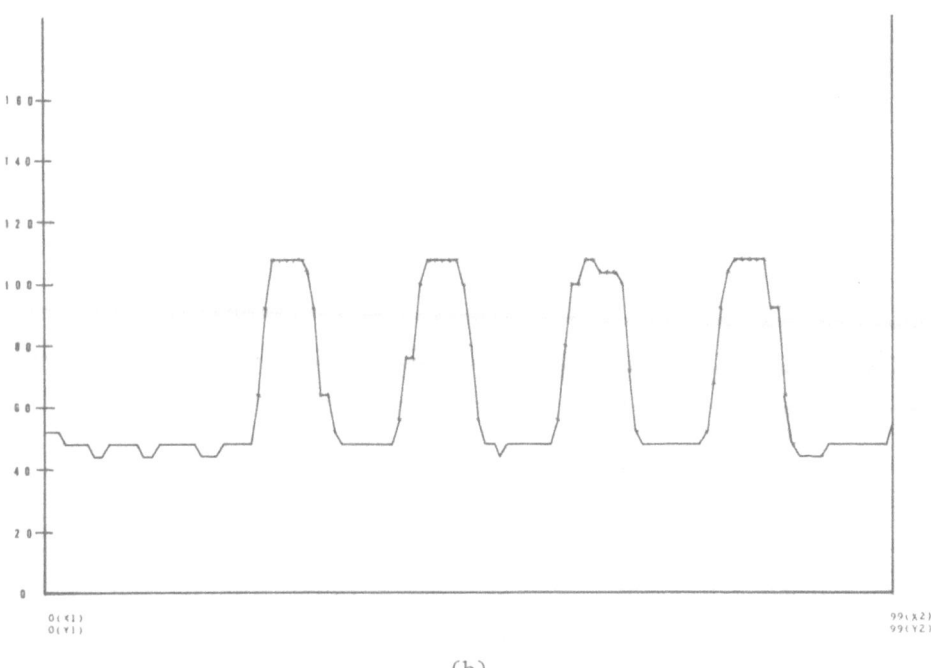

(b)

Fig. 3.8   (a) A blurred artificial pattern; the noisy artificial
           pattern of Fig. 3.5(a) is blurred by averaging over a
           5 × 5 neighborhood at each point in the picture

           (b) The cross section of (a) along the diagonal line
           from the upper left to the lower right corner

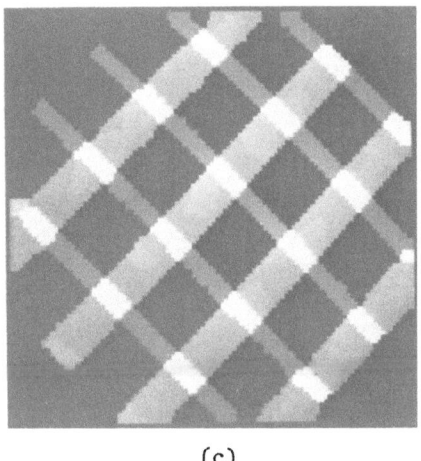

(c)

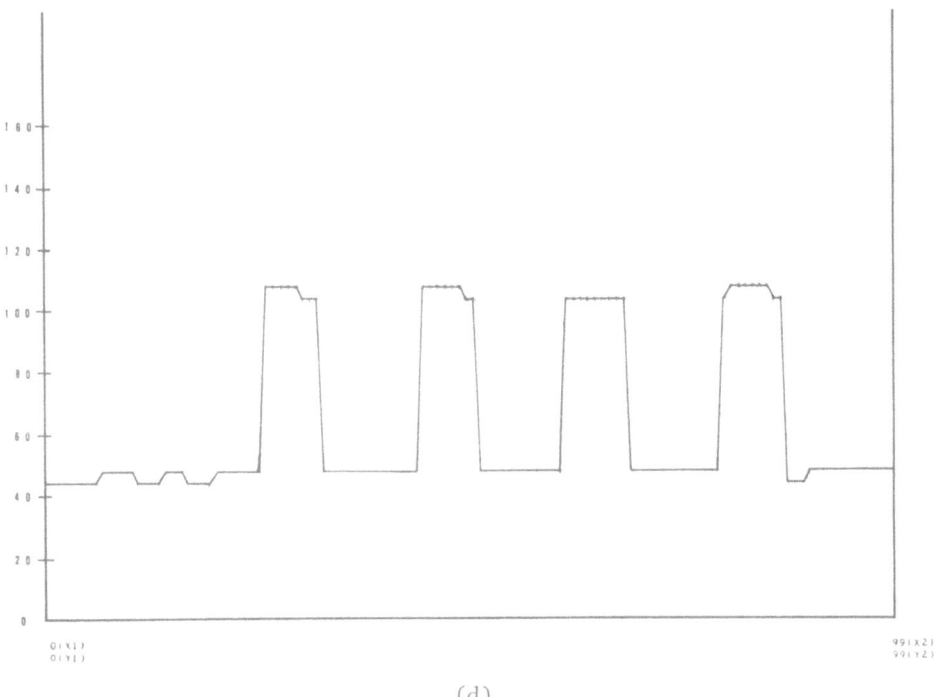

(d)

(c) Result of the edge-preserving smoothing of (a) after ten iterations

(d) The cross section of (c) along the diagonal line

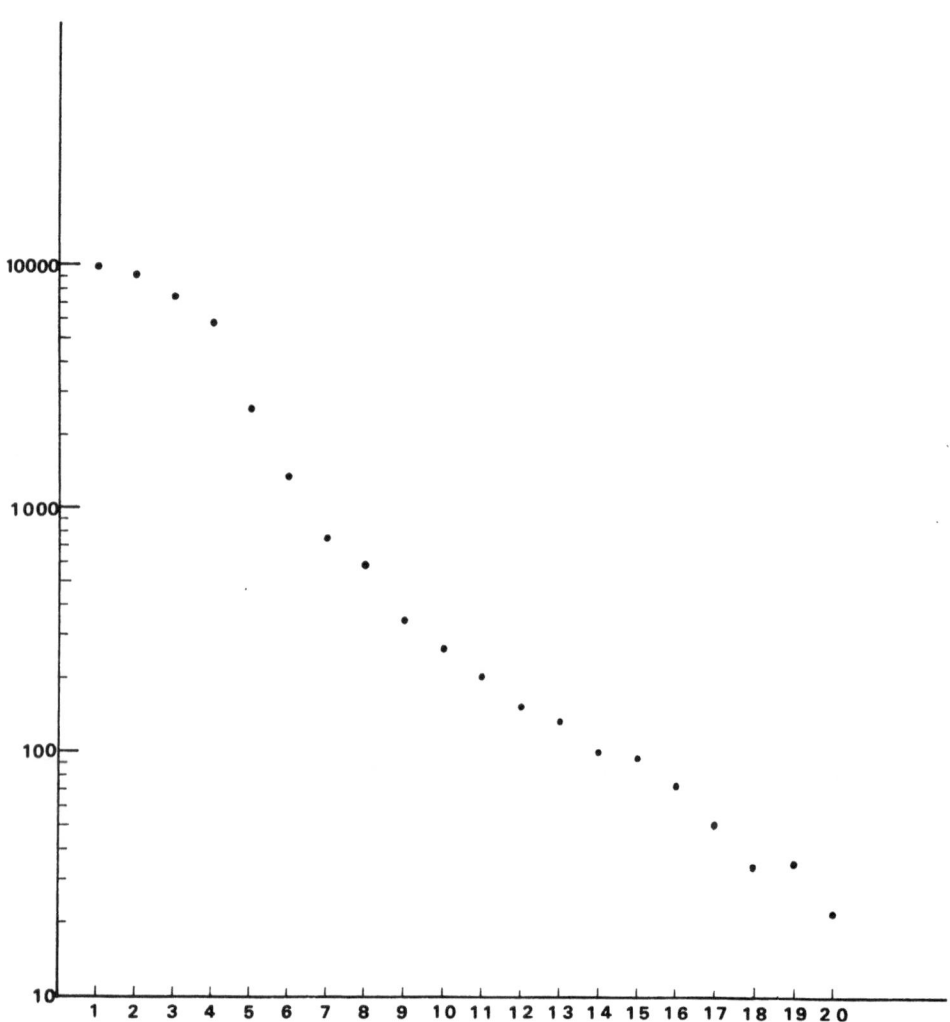

Fig. 3.9 Convergence of the edge-preserving smoothing of the noisy
          artificial pattern of Fig. 3.5(c); the horizontal axis
          shows the number of points whose gray levels are changed
          by smoothing (on a log scale)

(a)

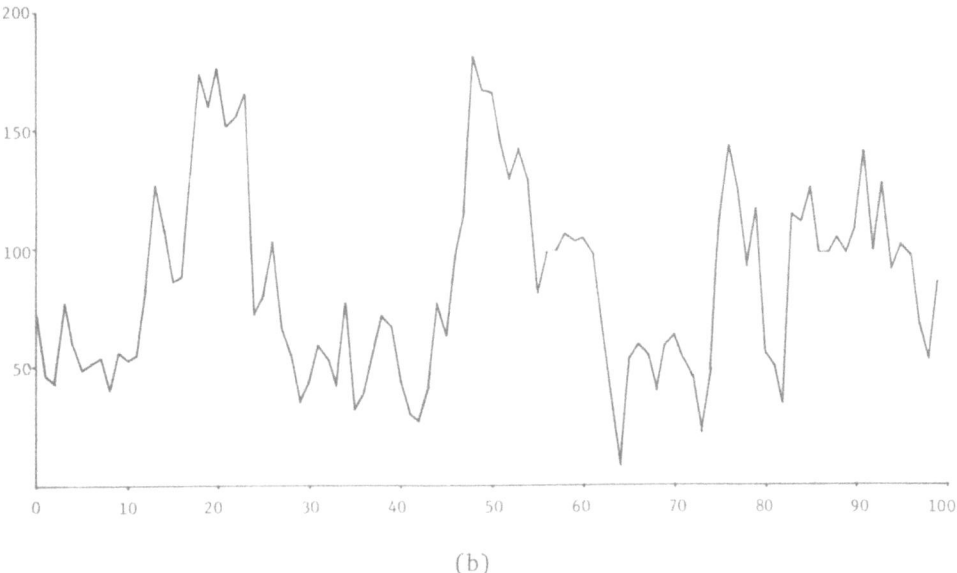

(b)

Fig. 3.10 (a) A noisy artificial pattern (the same as Fig. 3.5(c))
(b) The cross section of (a) along a horizontal line

(c)

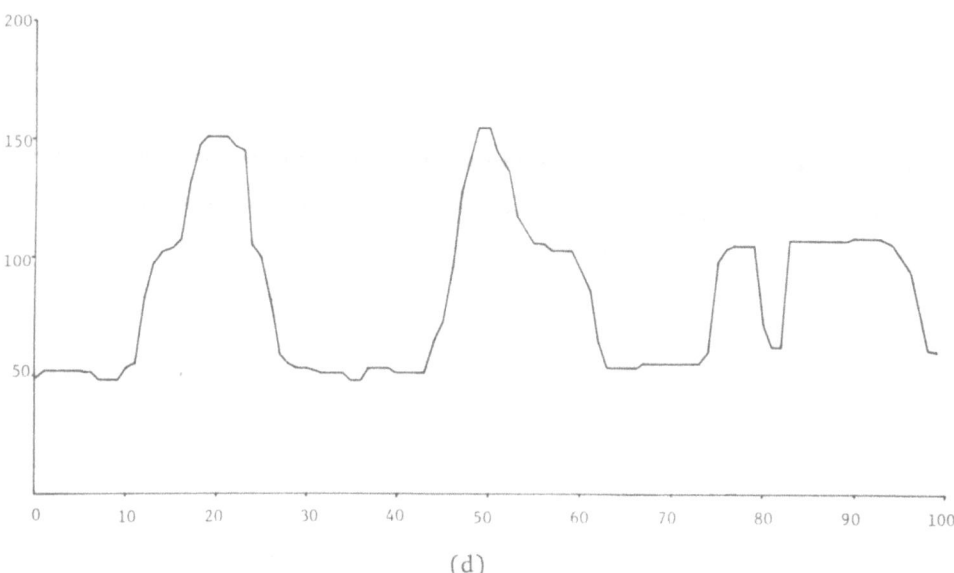

(d)

Fig. 3.10   (c) Result of median filtering after ten iterations
contd.
            (d) The cross section of (c) along the horizontal line

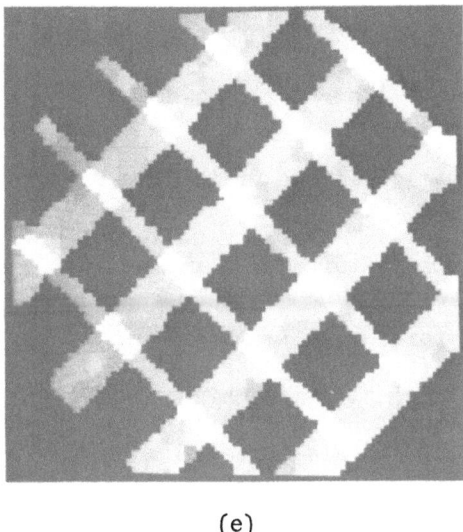

(e)

(f)

(e) Result of edge-preserving smoothing after ten iterations

(f) The cross section of (e) along the horizontal line

or edges from complex natural scenes. Our system for the structural analysis of aerial photographs incorporates this smoothing as the preprocessing for segmentation. Experimental results of applying this smoothing to complex aerial photographs will be presented in Section 4.2.

3.2 A Measure of Elongatedness of a Region

Shape discrimination is one of the central problems in pattern recognition and picture processing, and it has received much attention by many researchers. A variety of techniques have been proposed to measure shape features of a region (for a survey of shape analysis, see [57]).

Fourier descriptors which transform a boundary of a region into a set of Fourier coefficients are said to be "information preserving" because if sufficiently many higher order coefficients are evaluated, one can reproduce an original two-dimensional shape to any degree of detail. In this sense, Fourier descriptors are quite useful to discriminate among silhouettes of objects, and have been used in many applications.

When we identify objects in a scene, however, we do not utilize such mathematical terms nor can we imagine the shapes from them. We describe the shapes of objects in terms of such properties as rectangularity, triangularity, elongatedness, circularity, curvature, and so on. Although these properties of shape are not information preserving, they can be very useful for embedding our own knowledge about shapes of objects into computer programs, because it is usually expressed by just such properties: for example, houses are rectangular, roads are very elongated, and so on.

As mentioned before, since shape features are not sensitive to the photographic conditions under which aerial photographs are taken, we can rely much more on them to locate objects on the ground surface. Thus, our system incorporates many programs which calculate various shape features of regions. Some of them are used to extract characteristic regions, and others to recognize objects. Among these features, elongatedness plays an especially important role in locating elongated objects such as roads, rivers, and railroads. As these objects have a distinguishing shape feature, namely elongatedness, we will be able to extract them rather easily if the elongatedness of a region is correctly measured.

In this section, we will describe a sophisticated algorithm to measure the elongatedness of a region.

3.2.1. Elongatedness of a Region

The simplest way to measure the elongatedness of a region is to take the ratio between the length and width of its minimum bounding rectangle, as shown in Fig. 3.11. (For the minimum bounding rectangle of a region, see Section 4.4.) As one can easily imagine, however, it does not show the correct elongatedness for curved regions.

Another way to measure the elongatedness which works well even for curved regions is as follows:

Step 1: If a region to be measured has holes, fill them in. (Since to define the elongatedness of a region with holes is very difficult, we calculate the elongatedness from its outer boundary by neglecting the holes.)

Step 2: Shrink the region until it vanishes. Let d denote the number of shrinking steps for the region to disappear.

Step 3: Take $S/4d^2$ as the elongatedness of the region, where S denotes the area of the region.

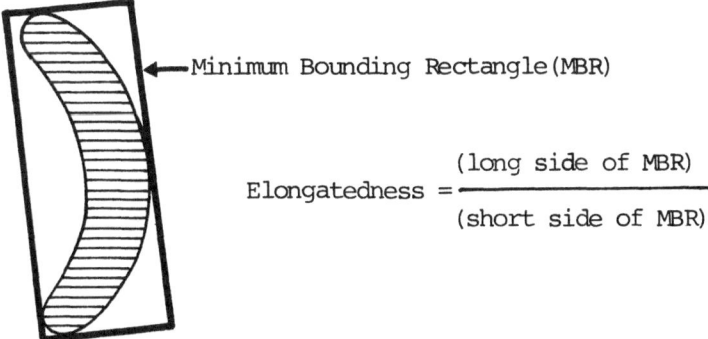

Minimum Bounding Rectangle (MBR)

$$\text{Elongatedness} = \frac{(\text{long side of MBR})}{(\text{short side of MBR})}$$

Fig. 3.11 Elongatedness calculated from the minimum bounding rectangle; the minimum bounding rectangle of a region is the minimum-area rectangle which encases the region

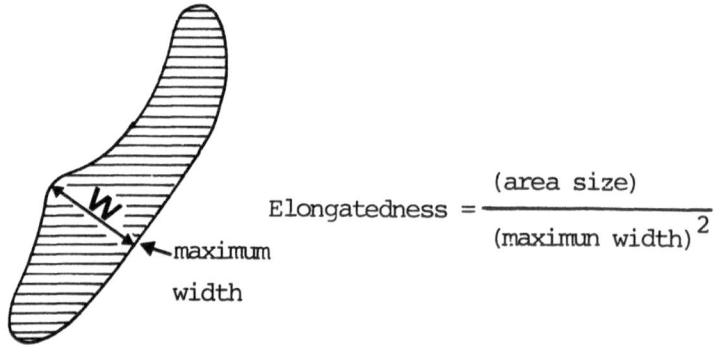

Fig. 3.12 Elongatedness calculated by using a shrinking opera-
           tion; this method takes the maximum width of the
           region as the width for calculating the elongatedness

Though this method is certainly able to measure the elongatedness
of a curved region, the width of the region calculated by this
method, *i.e.*, $2d$, corresponds to the maximum width of the region
(Fig. 3.12). Therefore the elongatedness tends to be underesti-
mated.

The algorithm we have developed first extracts the longest
path on the skeleton of a region, and then measures the width at
each point on the longest path. The elongatedness is defined as
the ratio between the average width of the region and the length
of the longest path. As elongated objects in aerial photographs
tend to be segmented into several regions because of small objects
on them such as shadows, cars, and bridges, the longest paths are
also very useful for connecting these segmented regions to form
the original unoccluded regions. Of course, we can define the
elongatedness as L^2/S without measuring the width of the region,
where L denotes the length of the longest path on the skeleton
and S the area of region. However the measurement of the change
of the width as well as both the longest path and the elongatedness of a region are quite valuable for recognizing elongated
objects. For example, we can use the knowledge that a road should
have a constant width. (The detailed algorithm to recognise elongated objects will be described in Section 6.3.)

3.2.2. Longest-Path Detection Algorithm

This algorithm extracts the axis of a region (the longest path) by pruning small branches of the skeleton of the region. Since it assumes that a region is "4-simply-connected" (connected in the meaning of 4-adjacency and has no holes in it), one should fill in holes before applying this algorithm.

Longest-Path Detection Algorithm

Step 1: Calculate the 4-connected skeleton of a region by thinning.

Step 2: Let $\vec{r}_i$ ($i = 1, \ldots N$) denote the positions of points on the skeleton.

Define the label of the point r_i, $\mathrm{LAB}\,(\vec{r}_i)$, as follows.

$\mathrm{LAB}(\vec{r}_i)$ = the number of points on the skeleton in the 4-neighborhood of $\vec{r}_i$.

That is, $\mathrm{LAB}(\vec{r}_i) = 1$: end point
2 : connecting point
3 : branching point
4 : crossing point.
The label of points which are not on the skeleton is set to 0.

Step 3: At the special pattern shown in Fig. 3.13(a), change the labels of the four branching points from 3 to 5 (Fig. 3.13(b)).

Step 4: Let $\vec{p}_j$ ($j = 1, \ldots, n$) denote the positions of end points.

Set $\mathrm{LAB}(\vec{p}_j) = -j$, $S_j = 0$ for all $j = 1, \ldots, n$
Set $\mathrm{ACTIVE} = n$ and $\ell = 1$

Step 5: If $\mathrm{ACTIVE} = 2$, then go to Step 9.

Step 6: If $S_\ell = 1$, then go to Step 8.

Else find $\vec{\delta}_k$ such that $L = \mathrm{LAB}(\vec{p}_\ell + \vec{\delta}_k) > 0$,

where $k = 1, 2, 3, 4$, and $\vec{\delta}_1 = (1,0)$, $\vec{\delta}_2 = (-1,0)$, $\vec{\delta}_3 = (0,1)$, $\vec{\delta}_4 = (0,-1)$.

Define the label of the point r_i, $\mathrm{LAB}\,(\vec{r}_i)$, as follows.

(It is evident from the algorithm that there exists only one $\vec{\delta}_k$ that satisfies the condition and that $L \neq 1$.)

Step 7: If $L = 2$, then $\text{LAB}(\vec{p}_\ell + \vec{\delta}_k) = -\ell$, $\vec{p}_\ell = \vec{p}_\ell + \vec{\delta}_k$.

If $L = 3$, then $\text{LAB}(\vec{p}_\ell + \vec{\delta}_k) = 2$, $S_\ell = 1$,
$\quad$ ACTIVE = ACTIVE - 1.

If $L = 4$, then $\text{LAB}(\vec{p}_\ell + \vec{\delta}_k) = 3$, $S_\ell = 1$,
$\quad$ ACTIVE = ACTIVE - 1.

If $L = 5$, then $\text{LAB}(\vec{p}_\ell = \vec{\delta}k = -\ell$, $S_\ell = 1$,
$\quad$ ACTIVE = ACTIVE - 1.

and if $\text{LAB}(\vec{p}_\ell + \vec{\delta}_k + \vec{\delta}_h) = 5$, then

$$\text{LAB}(\vec{p}_\ell + \vec{\delta}_k + \vec{\delta}_h) = 2,$$

if $\text{LAB}(\vec{p}_\ell + \vec{\delta}_k + \vec{\sigma}_m) = 5$, then

$$\text{LAB}(\vec{p}_\ell + \vec{\delta}_k + \vec{\sigma}_m) = 3,$$

where $k, m, h = 1, 2, 3, 4$ and $\vec{\sigma}_1 = (1,1)$,

$$\vec{\sigma}_2 = (1,-1), \; \vec{\sigma}_3 = (-1,1), \; \vec{\sigma}_4 = (-1,-1).$$

Step 8: $\ell = \ell + 1$. If $\ell > n$ (the number of end points), then $\ell = \ell - n$. Go to Step 5.

Step 9: Let a and b denote the indices of end points such that $S_a = 0$, $S_b = 0$. The 8-connected path connecting those points whose labels are $-a$, $-b$ or 2 becomes the longest path on the skeleton.

Note: The reason why we do not apply an 8-connected thinning from the first is that it is very difficult to trace an 8-connected skeleton since it sometimes has very complex patterns around crossing and branching points. In a 4-connected skeleton there exists only one special pattern shown in Fig. 3.13(a).

This algorithm traces the points on the skeleton from each end point simultaneously, and if it reaches a branching or crossing point, it stops proceeding beyond that point and becomes inactive (*i.e.*, $S_\ell = 1$). When the number of active points ($S_\ell = 0$) becomes 2, we have the longest path on the skeleton.

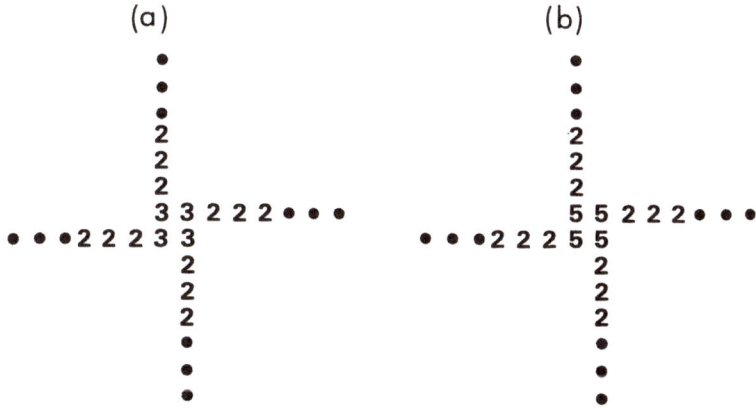

Fig. 3.13   (a) Special pattern of crossing points

(b) Relabeling of four crossing points

Figure 3.14 shows several examples of the longest paths of regions.

3.2.3.   Calculation of the Elongatedness

After extracting the longest path on the skeleton, the elongatedness of a region is calculated as follows.

Step 1:   Let $\vec{r}_i$ ($i = 1, \dots N$) denote the position of the points on the longest path. Calculate the width of the region at $\vec{r}_i$, w_i,

$$w_i = n_1 + \sqrt{2}\, n_2 \ ,$$

where n_1 and n_2 denote the number of steps along the horizontal or vertical line and the diagonal line, respectively, when we trace the points in the region along the straight line which passes through $\vec{r}_i$ and is perpendicular to $\vec{r}_{i-1} - \vec{r}_{i+1}$.

Step 2:   Calculate the average width of the region, W:

$$W = \frac{1}{N-2} \sum_{i=2}^{N-1} W_i \; .$$

Step 3:   Elongatedness, ELONG2, is defined as
ELONG2 = (length of the longest path)$/W$.

It is evident from the above processes that ELONG2 can measure the elongatedness of a region correctly even if it is very curved. The numbers shown in Fig. 3.14 denote "ELONG's" of the various regions.

Although this algorithm is very complex compared to the methods which use the minimum bounding rectangle or the shrinking operation, it can give the axis of a region and can measure the change of width, which are very useful for the recognition of elongated objects.

3.3.   A Structural Description of Regularly Arranged Patterns

3.3.1.   Introduction

Texture gives an important feature to characterize and discriminate regions. Because of the importance of textural properties in various applications, such as remote sensing and biomedical image analysis, many methods have been developed to extract textural features. They can be classified into two categories: statistical and structural methods.

In the statistical approach, texture is characterized by a set of statistics which are extracted from a large ensemble of local properties representing interpixel relationships. The gray level co-occurrence matrix proposed by Haralick [29] is one of the most widely used methods. Generally speaking, the statistical methods are useful for very fine textures which do not contain any obvious "texture elements" or regular spatial arrangements.

On the other hand, the structural methods are based on the model that textures are made of a set of elements arranged according to some regular placement rules. Thus, to describe the structure of texture, we have to characterize the elements and the placement rules. (For a survey of various statistical and structural approaches to texture analysis, see [31].)

We can find many examples of textures in an aerial photograph: a fine texture of low contrast in a grassland, a coarse texture of high contrast in a forest area, and a regular texture of large elements in a residential area. While the statistical

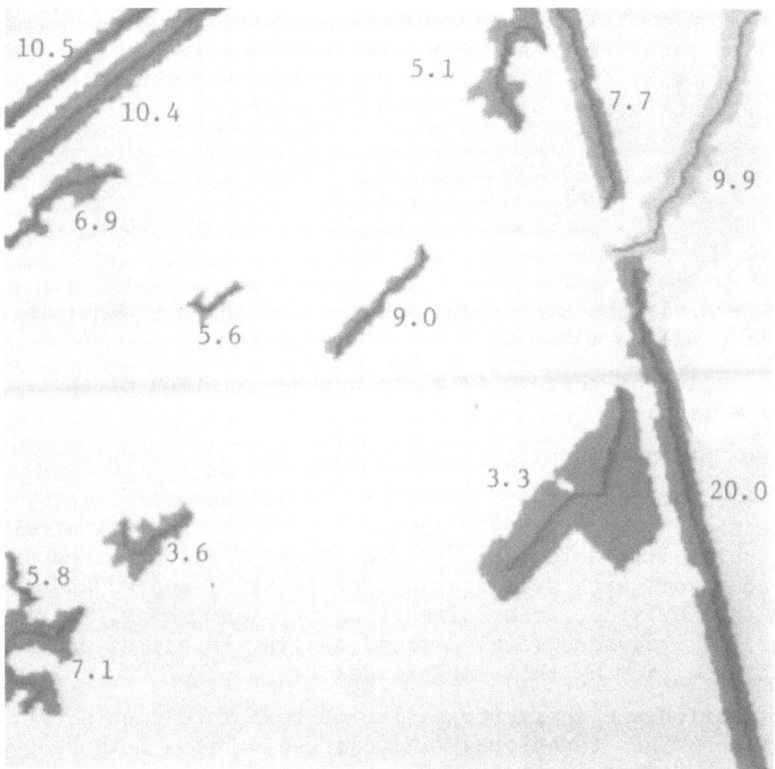

Fig. 3.14 Examples of the longest paths of regions; the numbers
denote the elongatedness of each region

approach may work well for characterizing natural textures in a
grassland and a forest area, the structural approach is more suitable for the artificial texture in a residential area which is composed of regularly arranged houses..

In this section we will present a new method to describe spatial relationships in regularly arranged patterns using relative
vectors between elements. We assume that texture elements have
already been extracted from a picture and classified into some
categories based on their properties. Thus, we are given a set of
triplets, (x_i, y_i, c_i) as input data, where c_i denotes a category
for the i-th element, and (x_i, y_i) denotes its position.

In our method, the spatial arrangement of elements in each
category is first described individually, and then a set of

descriptions for different categories of elements are combined to represent the overall arrangement of elements. In what follows, we shall present first the method for describing spatial arrangements of elements of a single category, and then the method for combining descriptions of different categories.

3.3.2. Describing Spatial Relationships Using Relative Vectors Among Elements

In general, the structural analysis of texture consists of two levels of processes:

(1) Extraction of texture elements and description of their properties.

(2) Description of spatial relationships among elements.

An element consists of a set of pixels characterized by a set of properties. The simplest element is a pixel and its attribute is a gray level. Tomita, Shirai, and Tsuji [77] extracted sets of connected pixels with similar gray levels as elements, and characterized them by size, area, directionality, and shape. Carlton and Mitchell [14] regarded local extrema of gray levels as elements and characterized them by their magnitudes.

Extraction and characterization of texture elements by themselves are not new techniques. We can use various methods of segmentation and shape analysis. How to describe spatial relationships among elements is a central problem in the structural analysis of texture, and requires intensive research.

Many methods have been proposed to describe placement rules among elements: a method using density of elements [78], a graphlike language to describe arrangements of lines and polygons [15], a tree grammer to express the spatial structure of pixels in an $N \times N$ window [45], and so on. Zucker [86] proposed a model of texture where an observed texture is considered to be generated as a distorted version of an ideal texture which is represented by a regular graph. (A regular graph is a graph in which every node is connected to neighboring nodes in an identical way.) The density of elements, however, is too simple a feature to be used for the description of spatial arrangements, and others are somewhat synthesis-oriented. We can not use them to extract spatial relationships among elements when we are given a picture of texture. We must develop an analysis-oriented method which can extract the description of spatial relationships from observed data.

Given the positions of N elements (E_i, i = 1, 2, ..., N), the most straightforward way to describe an arrangement among them is to use relative vectors between elements (here we assume that all

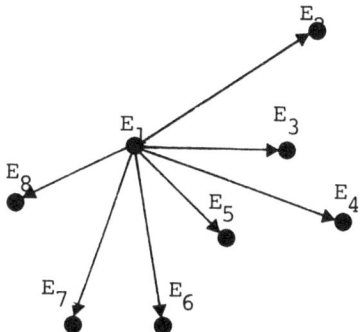

Fig. 3.15 Relative vectors for describing the structure of the
           arrangement

elements belong to the same category). That is, their spatial relationships can be exactly described by at most N-1 relative vectors, each of which defines a relative position between some fixed
element and one of the other elements (Fig. 3.15). Though a random
arrangement of elements requires N-1 vectors, the number of vectors
can be reduced if the arrangement is very regular. If the elements
are located on a two-dimensional lattice (in Fig. 3.16, the two
sides of the lattice need not necessarily cross at right angles),

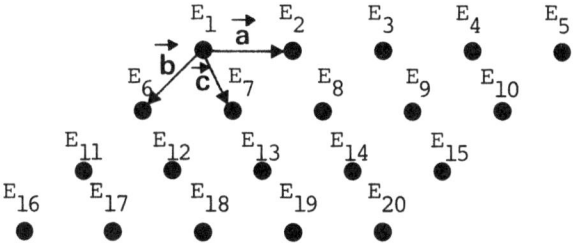

Fig. 3.16 A two-dimensional lattice arrangement

two vectors which generate the lattice, *i.e.*, $\vec{a}$ and $\vec{b}$, are sufficient for describing the spatial arrangement. Thus, the number of
vectors required to describe a spatial arrangement among N elements is confined to between 1 and N-1; if all elements are arranged on a straight line at equal intervals, only one vector is
sufficient.

From the above discussion it becomes clear that relative vectors among elements can be useful vocabularies with which we describe the structure of a spatial arrangement among elements. But
when we are going to extract the structure among given elements,
we have to solve the following problems. Since elements are not
always located at completely regular positions, how can the fluctuations in their positions be removed? Since there are $N(N$-1$)/2$
relative vectors among elements and most of them are redundant,
how can we select vectors with which the spatial arrangement is to
be described? For example, in Fig. 3.16, $\vec{a}$ and $\vec{b}$, $\vec{a}$ and $\vec{c}$, or $\vec{b}$
and $\vec{c}$ can potentially generate the same lattice. Then we must
determine which pair of vectors is the most suitable for the description of the arrangement.

3.3.3. Extraction of Regularity Vectors

In our method we first define "regularity vectors" from a set
of relative vectors between elements, and then try to find the
"simplest" description of the spatial arrangement in terms of the
regularity vectors.

> DEFINITION: A regularity vector is defined as a relative
> vector which appears quite often in an arrangement of
> elements.

In Fig. 3.17, $\vec{a}$, $\vec{b}$, $\vec{c}$, etc., are regularity vectors, while a
relative vector between elements E_1 and E_{27} is not as it occurs
only once in this arrangement. (All arrangements of elements we
consider are those in the "finite" two-dimensional space.)

Since we have no simple way of describing random arrangements,
we assume that the arrangements under consideration are regularly
repetitive patterns. (Some discussions on non-repetitive arrangements will be given in Section 3.3.6.) There are many regularity
vectors in an arrangement of elements. All spatial relationships
among elements necessary for description are represented in the
set of regularity vectors even if the elements are arranged in a
hierarchical way. For example, it seems to be natural to describe
the arrangement in Fig. 3.17 as having a two-layered structure;
one describes a triangular arrangement among three neighboring

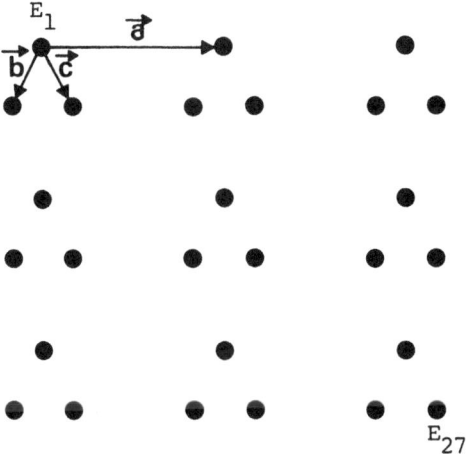

Fig. 3.17 A hierarchical arrangement of elements; $\vec{a}$, $\vec{b}$ and $\vec{c}$ are
regularity vectors while a relative vector connecting
E_1 with E_{27} is not

elements, and the other a two-dimensional lattice of local triangles. Since the vectors required to describe the triangle and
the lattice often appear in this arrangement, they are always extracted as regularity vectors.

The first step of processing for the description of the spatial arrangement of elements is to extract regularity vectors from
a set of relative vectors between elements. But the positions of
elements, which have been calculated by analyzing a picture of
texture, often fluctuate around ideal positions because of noise,
so that there exist no relative vectors which coincide completely
with others. Therefore, we apply a clustering technique in the
two-dimensional "vector space", where a point (x, y) denotes a
vector whose components are x and y (Fig. 3.18). We define regularity vectors as central points of clusters with large population.
As a result, all fluctuations of positions are removed. We have a
set of regularity vectors $\{v_1, v_2, \ldots, \vec{v}_n\}$ and n sets of elements,
each of which is composed of pairs of elements connected by a regularity vector. For example, in Fig. 3.16, we have $\vec{a}$ as a regularity vector and a set of pairs of elements whose spatial relationships are represented by $\vec{a}$, $\{(E_1, E_2), (E_2, E_3), (E_3, E_4),$

(E_4, E_5), $(E_6 \ E_7)$, ..., $(E_{19}, E_{20})\}$. To describe the structure
of the arrangement among elements, we only have to perform symbolic
manipulations on a set of primitive symbols $\{E_1, E_2, ..., E_n\}$, among
which n types of binary relations $\{v_1, v_2, ..., v_n\}$ are defined.

The clustering procedure to extract regularity vectors is as
follows:

Step 1: Plot $N(N-1)/2$ relative vectors between all pairs
of elements into the vector space, and memorize
starting and ending elements for each relative
vector.

Step 2: Apply the furthest-neighbor clustering technique
using the absolute value of the shortest relative vector as the threshold.

(I) Calculate distances between all pairs of different clusters in the vector space (initially
each cluster consists of a relative vector),
where the distance $D(i, j)$ between i-th and j-th
clusters is defined by

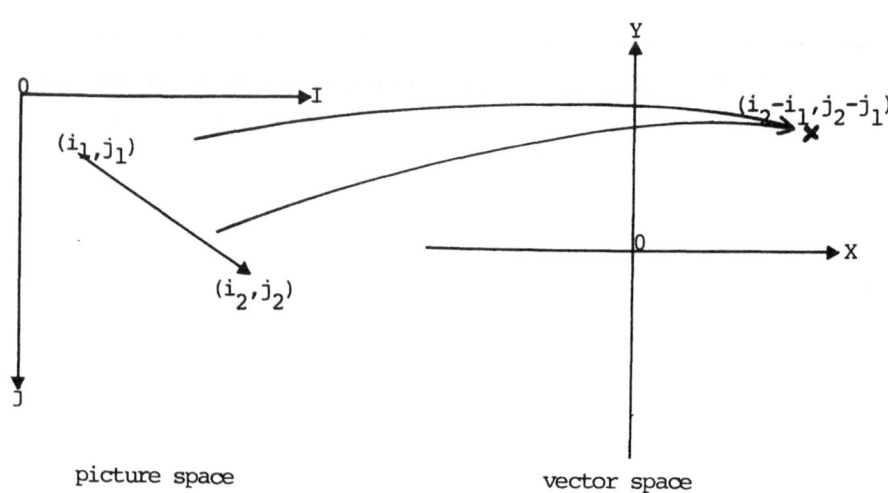

picture space   vector space

Fig. 3.18 Transformation from the picture space to the vector space

$$D(i,\ j) = \max_{k,\ell}\left\{\min\left(\sqrt{(x_i^{\ k} - x_j^{\ \ell})^2 + (y_i^{\ k} - y_j^{\ \ell})^2},\right.\right.$$

$$\left.\left.\sqrt{(x_i^{\ k} + x_j^{\ \ell})^2 + (y_i^{\ k} + y_j^{\ \ell})^2}\right)\right\}, \qquad (1)$$

where $(x_i^{\ k},\ y_i^{\ k})$ and $(x_j^{\ \ell},\ y_j^{\ \ell})$, $(k = 1,\ \ldots,\ N_i$, $\ell = 1,\ \ldots,\ N_j)$ denote points (relative vectors) in i-th and j-th clusters respectively, and N_i and N_j denote the numbers of points in them, respectively.

(II)   Let $i*$ and $j*$ denote a pair of clusters whose distance takes the minimum value. If $D(i*,\ j*)$ is less than the absolute value of the shortest relative vector, then merge $i*$-th and $j*$-th clusters into one cluster and go to (I)

Step 3:   Reject those clusters whose populations are less than the threshold.

Step 4:   For each detected cluster exchange, if necessary, starting and ending elements of relative vectors included in it so that the y component of the center of gravity of the cluster becomes nonnegative. (First exchange starting and ending elements of relative vectors whose y components are negative. Then calculate the center of gravity of the cluster. If its y component is negative, exchange starting and ending elements of all relative vectors in the cluster. By this operation the y component of the center of gravity of the cluster becomes nonnegative.) Then, define the center of gravity as a regularity vector and list pairs of elements connected by relative vectors in the cluster.

By the furthest-neighbor clustering method, the distance between any pair of points in a resultant cluster cannot exceed the length of the shortest relative vector, and we can find the regularity vectors even if the elements fluctuate to some extent. Thus each regularity vector no longer denotes a real relative vector, but denotes symbolic relationships between pairs of elements.

Since two similar relative vectors are sometimes plotted in

opposite directions around the origin of the vector space, relative vectors of opposite directions are to be treated as the same. To handle the vectors of opposite directions as the same, we define the minimum between

$$\sqrt{(x_i^k - x_j^l)^2 + (y_i^k - y_j^l)^2} \text{ and } \sqrt{(x_i^k + x_j^l)^2 + (y_i^k + y_j^l)^2}$$

as a distance between (x_i^k, j_i^k) and (x_j^l, y_j^l). Then, by using this distance, we can merge two vectors even if they are pointed in opposite directions.

3.3.4. Estimating Locations of Missing Elements

In general, a picture of texture is corrupted by noise, and moreover the size of and the spaces between texture elements are very small. Therefore, it is very difficult to extract all elements from the picture correctly by ordinary picture processing methods, such as edge extraction and region growing. It often happens that some elements are left unextracted and some false elements are extracted. If most of the elements have been correctly extracted, we can estimate the positions of the missing elements by using the regularity vectors which have been calculated from the already extracted elements.

The algorithm for extracting missing elements is as follows:

Step 1: Calculate regularity vectors from the already extracted elements.

Step 2: Suppose that $\vec{v}$ denotes some regularity vector, then shift an element by $\vec{v}$ (or $-\vec{v}$). If there exist no elements at a shifted position, while another element is found at the position shifted by $2\vec{v}$ ($-2v$) (Fig. 3.19), examine the area carefully to see whether a new element can be found.

Step 3: Apply the above process to all combinations of elements and regularity vectors. If new elements are extracted, return to Step 1 and recalculate the regularity vectors.

This feedback procedure will allow us to extract missing elements which have been left unextracted at the first stage of the analysis. On the other hand, if some false elements have been

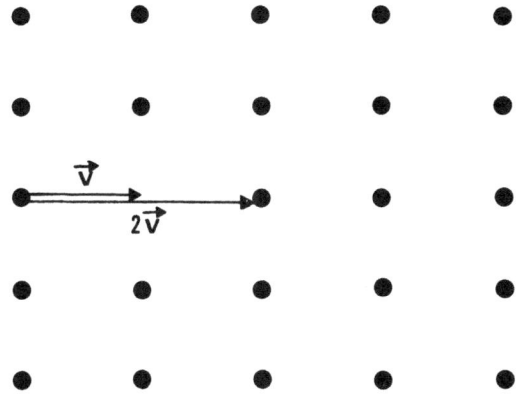

Fig. 3.19 Estimating locations of missing elements

extracted due to errors made by the picture-processing programs,
the clustering process for regularity vectors will reject them.
False elements usually appear at random in an arrangement. The
number of relative vectors similar to those from and to the false
elements is usually very small, so that the relative vectors from/to
the false elements are not recognized as regularity vectors. As a
result, no relationships by regularity vectors are defined over
false elements. For example, consider Fig. 3.20, where a false
element E' happens to be extracted by picture-processing programs.
As the position of E' is not correlated with the arrangement of
real elements, relative vectors connecting E' to the others appear
only once in this arrangement. Then, the clustering procedure re-
jects all relative vectors from and to E', so that E' is not considered as an element at the later stages of the analysis.

3.3.5. Describing Spatial Arrangements of Elements

We have a set of regularity vectors and the corresponding
sets of element pairs. We now proceed to the process of describing
the spatial arrangements of the elements by using this information.

As mentioned before, there exist many different ways of describing a given arrangement of elements in terms of regularity
vectors, so that we have to incorporate a criterion from which we
determine the most suitable description for a given arrangement.

It is natural to consider that the simplest description is the best.
In our method we have adopted the number of distinct regularity
vectors used for description as the measure of simplicity. We try
to describe the structure of the arrangement by using as small a
number of regularity vectors as possible. For this purpose, it
seems to be natural to select the regularity vectors which can represent long one-dimensional repetitive sequences of elements.

3.3.5.1. One-Dimensional Repetitive Pattern

If the arrangement under consideration is a one-dimensional repetitive pattern (Fig. 3.21), we can describe it with a single regularity vector $\vec{a}$ which connects any two neighboring elements. That
is, this arrangement can be described as a repetitive sequence of
elements defined by $\vec{a}$. Let us denote this sequence by $\vec{a}$ $[E_1, E_2,$
$\ldots, E_{10}]$. On the other hand, if we use $2\vec{a}$ as a relation between
elements ($2\vec{a}$ is also a regularity vector), the total set of elements is divided into two disjoint repetitive sequences, $\vec{a}$ $[E_1,$
$E_3, E_5, E_7, E_9]$ and $2\vec{a}$ $[E_2, E_4, E_6, E_8, E_{10}]$, and then we must describe the relation between these two sequences by $\vec{a}$. Comparing the
above two descriptions, it is clear that the former is simpler and
more suitable. Thus, the first step of describing the spatial
arrangement is as follows:

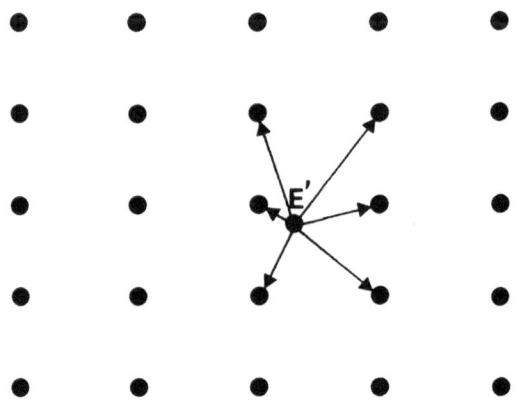

Fig. 3.20 Removing a false element (E') (see text)

Step 1:   Remove from a set of regularity vectors those
          vectors that are multiples of other regularity
          vectors. If we have a single vector left, the
          arrangement under consideration is a one-
          dimensional repetitive pattern, which can be
          described by that vector.

3.3.5.2. Two-Dimensional Lattice

Extending this idea to two-dimensional arrangements, the simplest arrangement, a lattice (Fig. 3.22(a)), can be described as a
repetitive sequence defined by $\vec{a}$ of a set of repetitive sequences
defined by $\vec{b}$, that is

$$\vec{a}\,[\vec{b}\,[E_1, E_2, E_3, E_4, E_5], \vec{b}\,[E_6, E_7, E_8, E_9, E_{10}],$$

$$b\,[E_{11}, E_{12}, E_{13}, E_{14}, E_{15}] \ .$$

Here we have assumed that $|\vec{a}| > |\vec{b}|$ and that the elements are
grouped into a set of repetitive sequences by a shorter vector and
then these sequences are grouped by a longer vector. (If $|\vec{a}| =$
$|\vec{b}|$, we first use the vector which makes a smaller angle with the
x-axis.)

Selecting $\vec{a}$ and $\vec{b}$ from a set of regularity vectors (there are
many other diagonal regularity vectors in this pattern) requires
some computation and is affected by the overall shape of the arrangement. For example, see Fig. 3.22(b), where the shape of the
arrangement is changed while keeping the same local spatial relationships between elements as Fig. 3.22(a). We can no longer describe this arrangement as a repetitive sequence defined by $\vec{a}$, but
we have to use $\vec{c}$ and describe it as a repetitive sequence defined

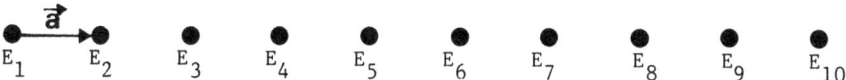

Fig. 3.21   One-dimensional repetitive pattern

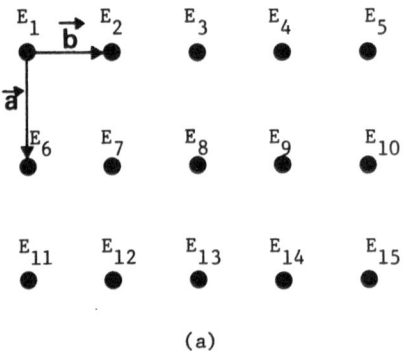

(a)

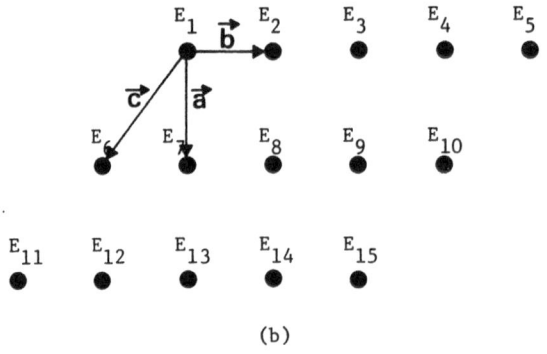

(b)

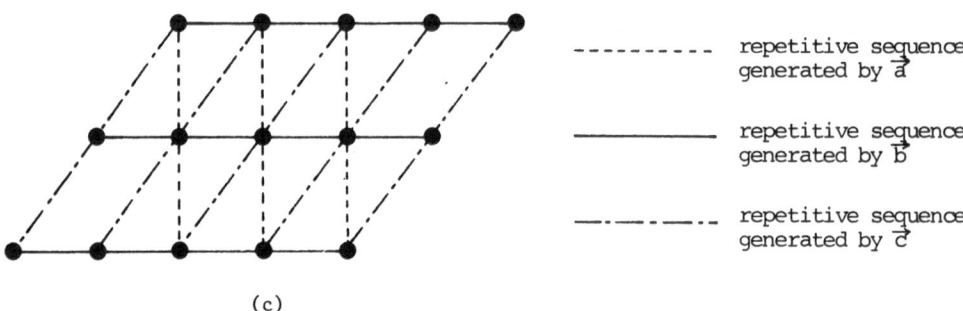

(c)

Fig. 3.22 Describing structures of two-dimensional lattices (see
          text)

by $\vec{c}$ of a set of repetitive sequences defined by $\vec{b}$, i.e.,

$$\vec{c}[\vec{b}[E_1, E_2, E_3, E_4, E_5], b[E_6, E_7, E_8, E_9, E_{10}],$$

$$\vec{b}[E_{11}, E_{12}, E_{13}, E_{14}, E_{15}]].$$

Thus, the shape of the arrangement as well as the local spatial relationships among the elements have crucial effects on the description (cf. in an "infinite" lattice two shortest linearly-independent
regularity vectors give the simplest description.)

Considering the above discussion, the following algorithm can
give the most suitable pair of regularity vectors to describe a
lattice.

Step 2:   Take any pair of vectors, $\vec{v}_i$ and $\vec{v}_j$ from a set

of regularity vectors. (Of course, some redundant vectors have already been removed by Step
1 in the previous section.)

Step 3:   Count the number of elements, N_{ij}, which are

contained in both of the repetitive sequences

defined simultaneously by v_i and v_j.

Note: A repetitive sequence defined by a vector $\vec{v}$ requires
at least three elements. That is, in Fig. 3.22(b),
E_4, E_5, E_6, E_{10}, E_{11}, and E_{12} are not contained in

any repetitive sequences defined by $\vec{a}$.

Step 4:   Select a pair of vectors as generating vectors
for the lattice such that N_{ij} is maximum among

all combinations of regularity vectors.

For the lattice shown in Fig. 3.22(b), if we use $\vec{a}$ and $\vec{b}$,,
only nine elements (Fig. 3.22(c)) are included in both of the one-
dimensional sequences defined by $\vec{a}$ and $\vec{b}$, and the rest of the elements are included in only repetitive sequences defined by $\vec{b}$.
Therefore, the description of the arrangement becomes complex if
we use these two vectors ($\vec{a}$ and $\vec{b}$). On the other hand, $\vec{b}$. and $\vec{c}$
can generate repetitive sequences which contain all 15 elements.
Thus, if the arrangement under consideration is really a lattice,
the above algorithm can find a pair of regularity vectors which
gives the simplest description.

3.3.5.3. Hierarchical Arrangement

Next, suppose that the arrangement is a two-dimensional repetitive pattern of some local arrangement of elements. For example,
see Fig. 3.23, where triangular arrangements of three elements are
arranged on a lattice. Then, we have $\vec{v}_1$, $\vec{v}_2$, $\vec{v}_3$, $\vec{v}_4$, etc. as regularity vectors. If we apply the above algorithm to this pattern,
we have $\vec{v}_3$ and $\vec{v}_4$ as the most suitable pair of regularity vectors,
because all elements are contained in both of the repetitive sequences defined by $\vec{v}_3$ and $\vec{v}_4$ at the same time. No other pairs of vectors
contain all elements in both of their repetitive sequences. Using
$\vec{v}_3$ and $\vec{v}_4$, the total set of elements is partitioned into three disjoint subsets each of which can be described as a lattice generated
by $\vec{v}_3$ and $\vec{v}_4$. Thus, we can describe this arrangement as an overlay
of three lattices, and the displacements between them are represented
by $\vec{v}_1$ and $\vec{v}_2$. (A method for selecting $\vec{v}_1$ and $\vec{v}_2$ is given below.)

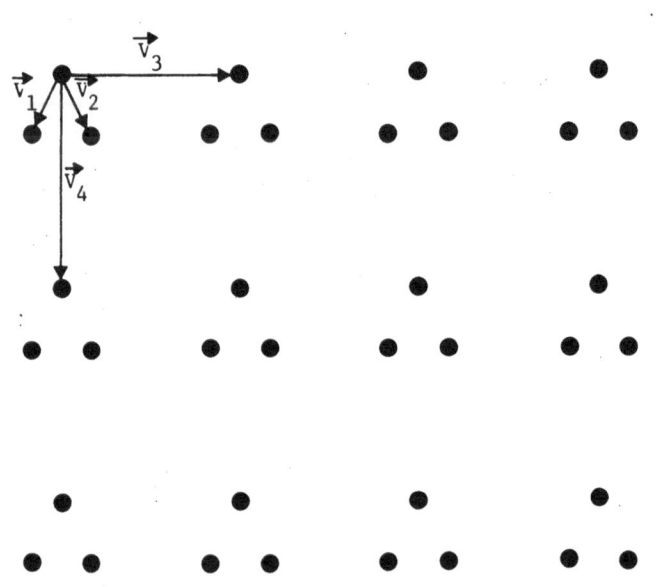

Fig. 3.23 A hierarchical arrangement of elements

Generally speaking, using the above algorithm, we can describe any hierarchical arrangement where a cluster of n elements
is placed on a lattice as an overlay of the same n lattices. The
displacements among these lattices can be determined by the following method.

Step 5: If a pair of regularity vectors selected by
Step 4 divides the total set of elements into
n disjoint subsets, then describe each subset
as a lattice generated by these vectors, and

(1) Find an upper-left element e_i for each lattice
$(i = 1, \ldots, n)$.

(2) Find the "upper-left-most" element $e*$ among
$\{e_1, e_2, \ldots, e_n\}$.

(3) Find regularity vectors connecting $e*$ and
e_i, and make them displacement vectors

between lattices.

By this method v_1 and v_2 are selected as the displacement
vectors between the three lattices in Fig. 3.23.

From the psychological point of view, this representation
sometimes does not coincide with human interpretation. That is,
we would like to represent the arrangement shown in Fig. 3.23 as
a lattice of local triangular arrangements of three elements.
This is because human beings tend to group neighboring elements
into a cluster and consider it as a high-level element. We do not
consider the arrangement in Fig. 3.24 as a lattice of triangles,
even though it obeys the same placement rule (in our sense) as in
Fig. 3.23, because the distances between the elements in the triangular arrangements are much longer than those in the lattices.

If one wants to get such a natural description (for human
beings) of a hierarchical arrangement, then compare the absolute
values between the regularity vectors generating the lattices and
those describing displacements between the lattices. If the
former are substantially larger than the latter, merge elements
located at the same position in the lattices into a cluster, and
regard it as an element. Then, the arrangement can be described
as a lattice of these new elements. In this case an element
itself will have a inner structure of its own. Although this processing can give a representation which coincides with the human
interpretation, it is not necessary for us, since we already have
a simple description.

So far, we have implicitly assumed that a unique pair of regularity vectors is extracted by Step 4, but in some cases this

is not the case. We have only one regularity vector for the arrangement of a one-dimensional repetitive pattern of local arrangements as shown in Fig. 3.25. Then, we describe it as an overlay
of one-dimensional repetitive sequences. (If needed, "Gestalt
clustering" may be performed first as in the case of Fig. 3.23.)

On the other hand, if a local arrangement of elements in a
hierarchical arrangement is itself a lattice, our algorithm gives
more than one pair of regularity vectors. For example, three and
six pairs of regularity vectors will be detected in the arrangements shown in Fig. 3.26 and 3.27, respectively, $i.e.$, $(\vec{v}_1, \vec{v}_2)$,
$(\vec{v}_1, \vec{v}_3)$, and $(\vec{v}_2, \vec{v}_3)$, and $(\vec{v}_1, \vec{v}_2)$, $(\vec{v}_1, \vec{v}_3)$, $(\vec{v}_1, \vec{v}_4)$, $(\vec{v}_2, \vec{v}_3)$
$(\vec{v}_2, \vec{v}_4)$, and $(\vec{v}_3, \vec{v}_4)$. These regular hierarchical arrangements
can be described as three or four levels of repetition of one-
dimensional repetitive sequences. That is,

> Step 6: If we have more than one pair of regularity
> vectors in Step 4 and by using any pair of re-
> gularity vectors the arrangement can be repre-
> sented as an overlay of lattices, we sort such
> regularity vectors according to their absolute
> values (from short to long). Let $\{\vec{v}_1, \vec{v}_2, \ldots,$
> $\vec{v}_\ell\}$ denote this sequence of selected regularity
> vectors. Then, we first find one-dimensional
> repetitive sequences defined by $\vec{v}_1$, and align
> these sequences in one-dimensional repetitive
> sequences using $\vec{v}_2$, and so on.

3.3.6. Discussion

The flow chart in Fig. 3.28 summarizes the method of describing structures of regularly arranged patterns which have been discussed in the previous sections. Here, we will discuss some ex-
tensions of our method to more complex and less regular arrangements.

When we are going to describe an arrangement composed of elements of different categories, we first describe the spatial arrangement among elements of the same category, respectively, and
then combine such "within-category" descriptions into an overall
description by using relative vectors between upper-left elements
of the descriptions. For example, see Fig. 3.29, where two classes
of elements are placed on lattices, respectively. The relation
between the two lattices (within-category description) can be

described with a relative vector which connects upper-left elements of the two lattices, i.e., $\vec{v}$. As these lattices have the same structure and the relative vector between them is much smaller than the vectors generating the lattices, we can consider each pair of neighboring elements of different categories as an element and describe the overall arrangement as a lattice of new elements. (Since regularity vectors generating two lattices of different categories have been calculated independently of each other, we need some procedures to recognize these two lattices as belonging to the same structure.)

Although all that our method can describe are regular repetitive patterns, it may be applicable to most of artifically generated textures, such as tileworks on pavements and walls, printed patterns on clothes, and so on. Fig. 3.30 shows eleven different designs for regular wall patterns (three regular tesselations and eight semi-regular tesselations), where the two-dimensional plane is covered with regular polygons without gaps or overlaps [20]. Though some of them seem to be very complex and difficult to describe, all of them can be described by our method as overlays of two-dimensional lattices of several kinds of elements. Polygons with different directionality (even if they have the same shape) are regarded as distinct kinds of elements. Thus, our method can analyze and describe a variety of regular arrangements.

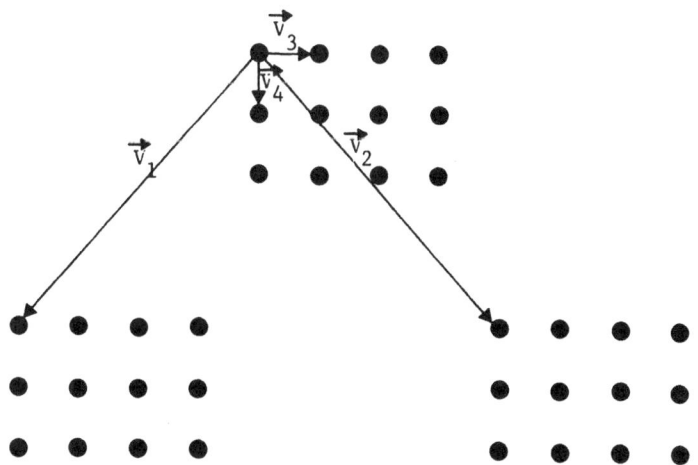

Fig. 3.24 A hierarchical arrangement with the same structure as that in Fig. 3.23

Fig. 3.25 A one-dimensional hierarchical arrangement

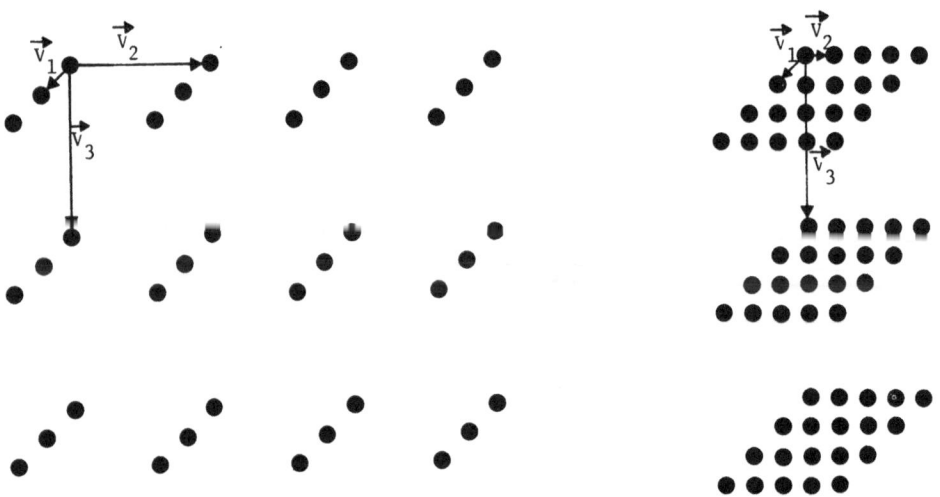

Fig. 3.26 Regular hierarchical arrangements

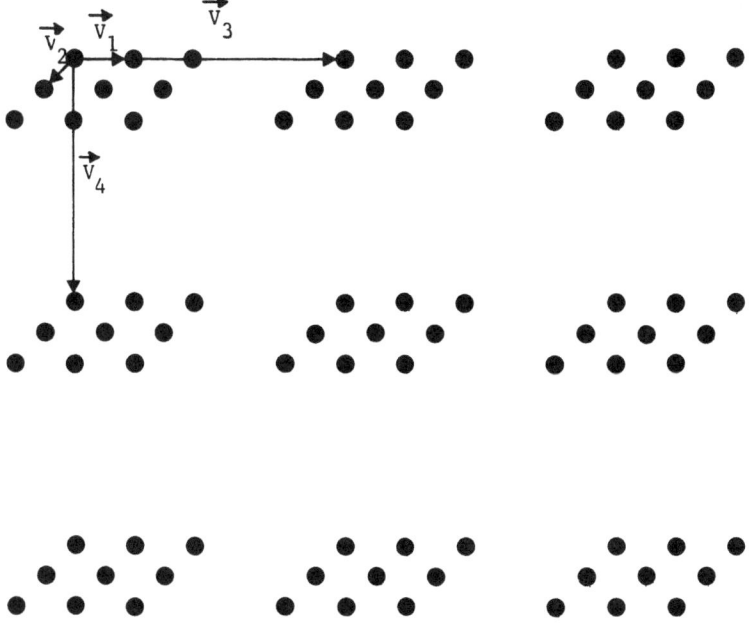

Fig. 3.27   Regular hierarchical arrangement

If arrangements under consideration are non-repetitive patterns as shown in Fig. 3.31, we can extract no regularity vectors
with which the overall structures of the arrangements can be described. However, we can characterize them by some statistics calculated from relative vectors. For example, if we plot relative
vectors between elements and their nearest neighbours into the
vector space, we can find that all points in the vector space are
located at the same distance from the origin (Fig. 3.31(a)) or are
aligned in the same direction (Fig. 3.31(b)). Thus, we can characterize such non-repetitive arrangements, but it is very difficult
to describe their structures in simple forms unless we have some
a priori knowledge about the arrangements.

Our method in substance decomposes arrangements into two-
dimensional lattices. Therefore, if some elements at the grid
points are randomly missing as shown in Fig. 3.32, it cannot give
any descriptions even if it can select the most suitable regularity
vectors ($\vec{v}_1$ and $\vec{v}_2$ for the arrangement in Fig. 3.32). But if the

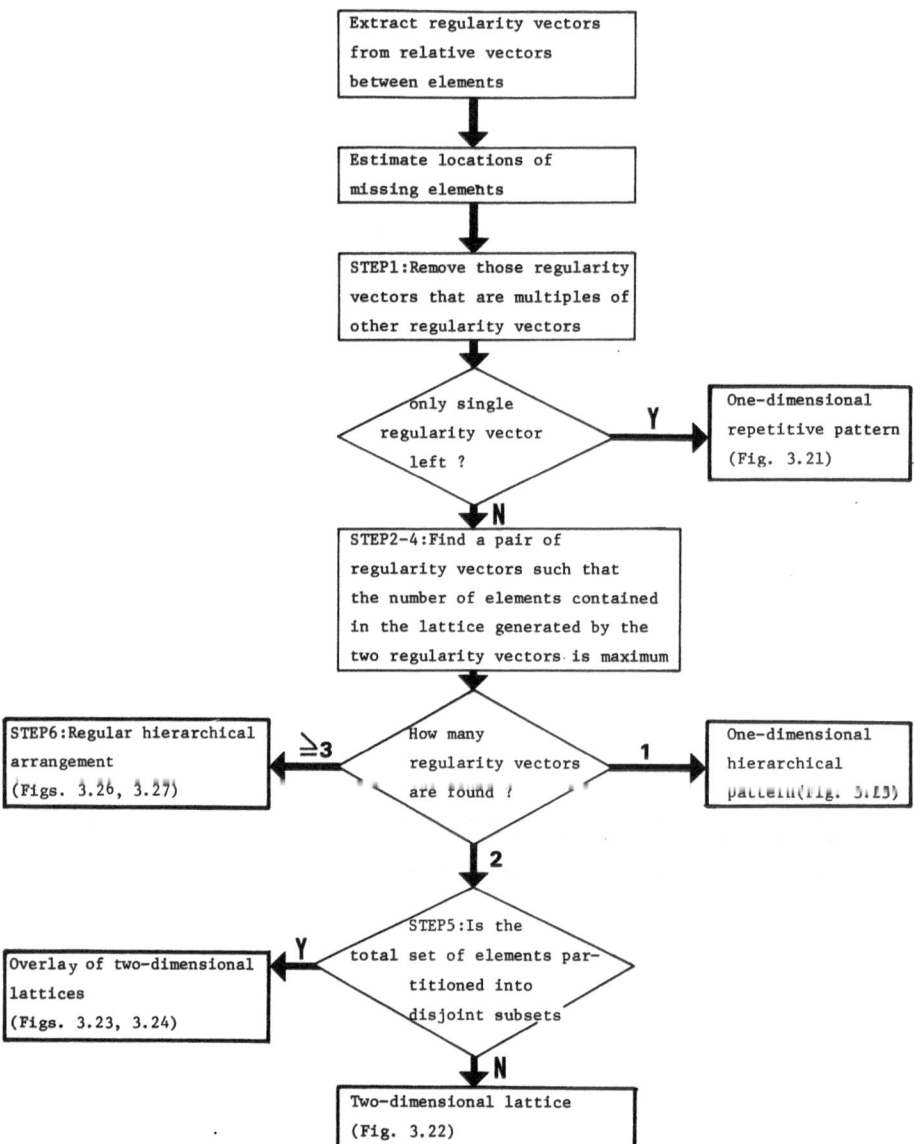

Fig. 3.28   Flow chart of describing structures of regularly
            arranged patterns

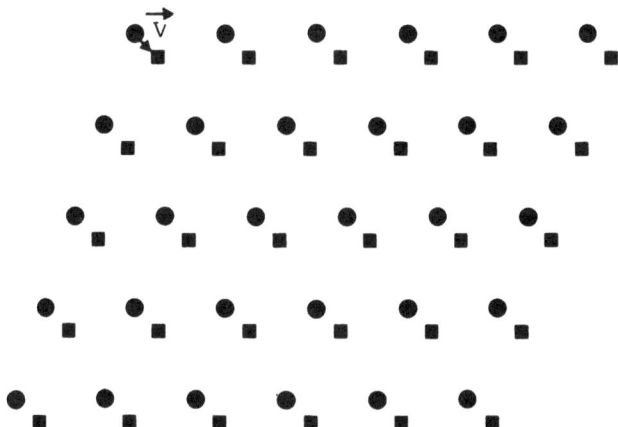

Fig. 3.29 Texture composed of two different kinds of elements

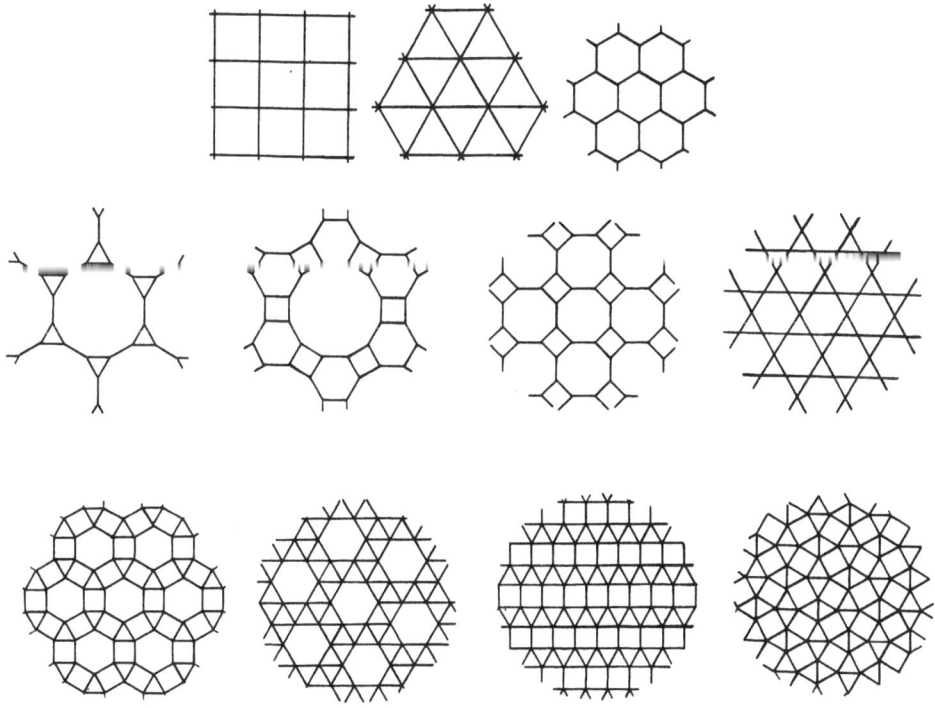

Fig. 3.30 Eleven regular wall patterns (from [20])

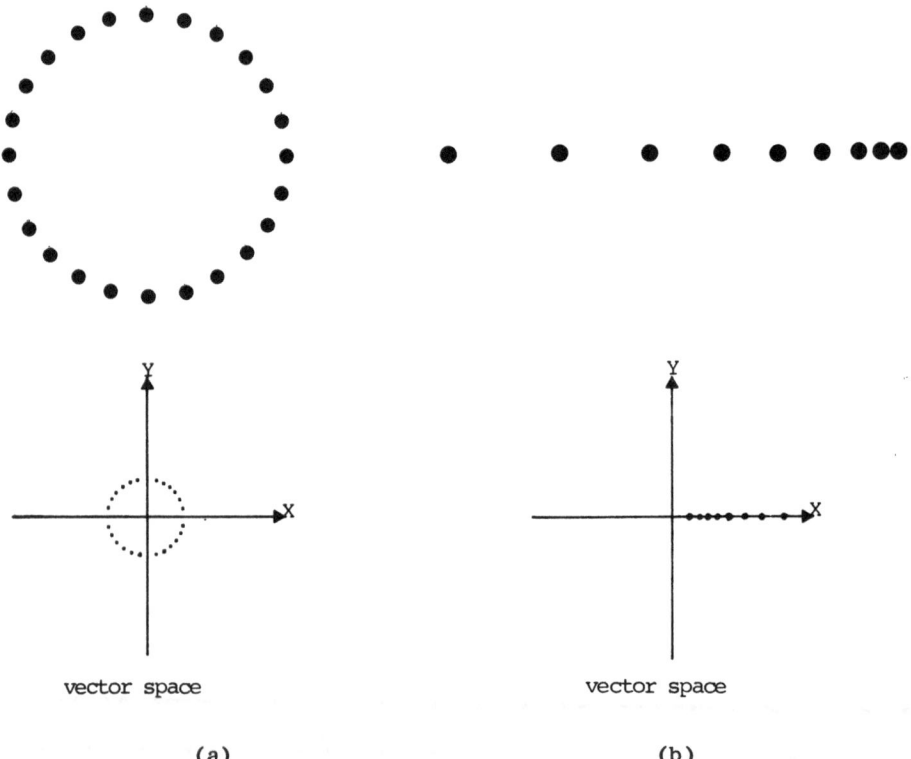

vector space   vector space

(a)   (b)

Fig. 3.31   Non-repetitive patterns; there exist no regularity
vectors in these patterns

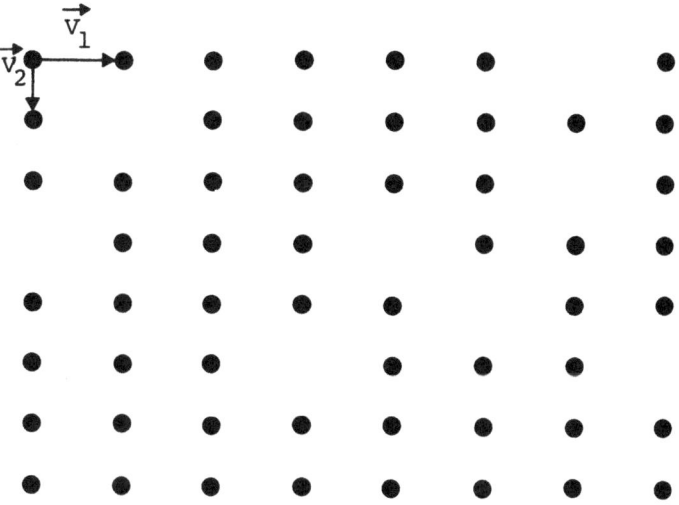

Fig. 3.32   Two-dimensional lattice with missing elements

missing elements themselves are arranged regularly, we can describe
the structure of the arrangement by generating pairs of "normal"
and "inverse" elements at positions of missing elements and can
describe it as an overlay of lattices of normal and inverse elements. For example, in Fig. 3.33, we generate pairs of ● (normal
element) and ○ (inverse element) at positions a, b, c, d, e, f,
g, h, and i, resulting in complete lattices of ● and ○ . We
assume that if a pair of ● and ○ are located at the same
place, we have nothing in the real two-dimensional space. Then,
the arrangement can be described as an overlay of the lattices of
● and ○ . The positions where normal and inverse elements are
to be generated can be found by the method described in Section
3.3.4. That is, if we cannot find any elements at the positions
estimated by using regularity vectors, we generate pairs of normal
and inverse elements at those positions. (If missing elements
come from errors in picture processing, we will be able to detect
elements at the estimated position.)

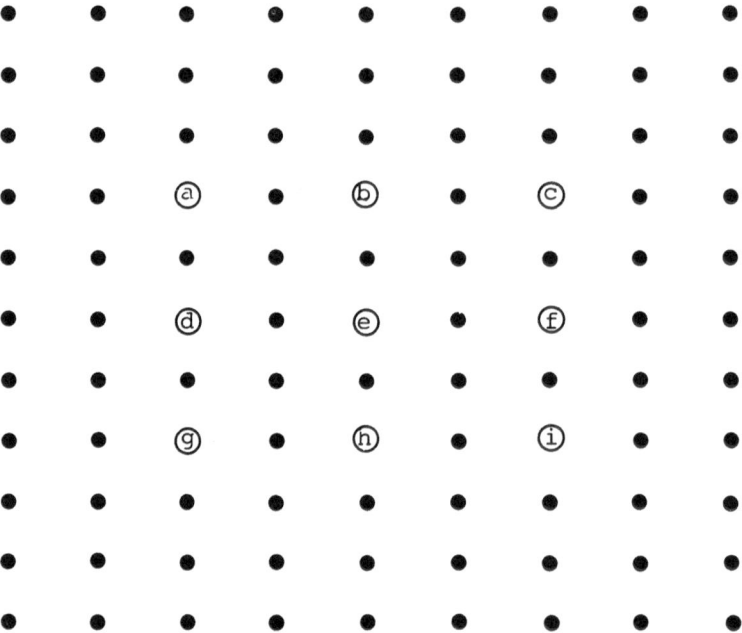

Fig. 3.33 Two dimensional lattice with missing elements; missing
elements themselves are regularly arranged

3.3.7. Conclusion

We have presented a method to describe structures of regularly arranged patterns as overlays of two-dimensional lattices.
We first extract regularity vectors from a set of relative vectors
between the elements. The regularity vectors can be used to locate
missing elements and remove erroneous elements, which is quite
helpful for the analysis of texture pictures because picture processing programs alone cannot perform a completely error-free extraction of elements.

Our method can construct the simplest descriptions of arrangements from a set of regularity vectors. It can be applied to a
very wide class of regularly arranged patterns.

4. STRUCTURING OF PICTURE DATA

4.1. Low-Level Processing in Image Understanding

Image understanding is a quite intellectual process which converts the vast quantities of sensory data (raw picture data) into a compact semantic description of the scene. Accordingly, it has to perform a wide spectrum of processing from signal processing to semantic processing. These different levels of processing interact in a very complex way; the signal processing removes unimportant sensory information such as noise, and the organization process transforms raw picture data into some well-structured form which can supply enough information to the knowledge-based syntactic and semantic processing. On the other hand, the syntactic and semantic processing gives the information about the global structure of the scene which guides the local analysis during picture data processing.

In the design of image understanding systems, how these diverse levels of processing can interact and communicate with each other is one of the difficult and important problems. Usually, the complexity of this interaction leads us to divide an image understanding system into two sub-processes: the low-level processing and the high-level processing. The low-level processing organizes the raw picture data into some symbolic structure allowing the high-level processing to devote itself to the interpretation of the scene. Thus, the first step of the processing in an image understanding system is to extract edges (lines) and various features of regions from the raw picture data. This process is usually called "segmentation". Many methods have been proposed for extracting regions and edges from a picture. (For a survey of various segmentation techniques see [62].) The information about regions and edges is transformed into a well-structured symbolic form containing various attributes of entities (regions and edges) and mutual relationships among them (adjacency, inclusion, etc.). Then the high-level processing works on this symbolic data to describe the structure of the scene.

While the high-level processing is, of course, a knowledge-
based process, there have been several discussions on whether or
not the low-level processing has to incorporate the knowledge of
the task domain to be analyzed. Many experiments on the analysis
of natural scenes have shown that simple knowledge-free picture
processing routines cannot give completely satisfactory results
and that some errors are inevitable; regions are sometimes over-
merged or oversplit, and edges are often divided into disjoint
segments. This has led some researchers to "knowledge-guided"
picture processing.

Tenenbaum and Barrow [75] and Yakimovsky and Feldman [84] introduced syntactic and semantic restrictions into the process of
region growing. The picture is first divided into a lot of patches
with constant color. Then these patches are merged with each other
based on the syntactic and semantic constraints as well as brightness and color information. They performed both segmentation of a
picture and semantic interpretation of a scene at a single processing stage. Shirai [69] developed a context-sensitive edge finder
which utilized the knowledge about the structure of the scene to
detect edges in pictures of polyhedra. The edge finder, which is
a simple picture processing routine, is applied to some local areas
in the picture which are specified by the knowledge of the edge
structure of polyhedra.

These studies have succeeded in incorporating some knowledge
about the scene into the low-level picture processing stage. However, the scenes that can be analyzed are restricted to some specific domains. That is, since they use quite specialized knowledge about a specific task domain at a very early stage of processing, the overall process, even the low-level picture processing, is heavily dependent on the task domain. Therefore, the
whole analysis process and the knowledge incorporated will have to
be newly specified when we want to analyze a scene from another
domain. This forces us to reconstruct and re-evaluate the whole
task-dependent system every time we undertake a new application.
In this sense, the idea of incorporating special "task-dependent"
knowledge into the low-level processing is useful for specialized
purposes, but cannot be a general methodology for image understanding of diverse classes of scenes.

Marr [46] and Zucker et al. [85] stood against the incorpora-
tion of syntactic and semantic knowledge into the early stages of
visual processing, and insisted that much nonsemantic processing
could be done on the picture data without knowing what is present
in a scene. Based on the observation that human visual systems
can extract some entity from a picture even if the scene is essentially nonsense, they proposed various methods for feature extraction and entity aggregation which worked quite well to isolate
figures from the background without recourse to semantics. Even
though these entities and figures are not perfect, they can be

quite useful as the processing units on which the high-level interpretation processes work.

   Considering the discussions mentioned above, Riseman and Arbib
[62] stated in their paper their attitude to the problem of the low-
level and high-level processing as follows.

   "We view the problem of image understanding as one of perfor-
ming initial segmentation via general procedures, feeding this low-
level output to a high-level system, and then allowing feedback
loops so that the interpretation processes can influence refined
segmentation. This allows semantic information to influence segmentation in a goal-oriented way without coupling all such knowledge directly into the low-level processing."

   Our system for the structural analysis of aerial photographs
is based on the same philosophy as posed in the above quotation.
The structuring processes for raw picture data consist of the following two steps:

   (1) Remove noise in an input picture and sharpen blurred
       edges in order to facilitate the subsequent segmen-
       tation.

   (2) Partition a picture into a set of homogeneous regions
       (elementary regions) by merging neighboring pixels
       with similar multispectral properties into one region.

All subsequent processes perform various operations based on these
elementary regions and their properties. They treat each elementary region as a unit for processing.

   This structuring process does not incorporate any specialized
knowledge about specific task domains and, in this regard, is applicable to any type of scene. However, the result of this processing,
as mentioned before, cannot be perfect and will contain some errors.
Our system tries to correct these errors by repairing the segmentation errors with the help of the feedback loop from the high-level
interpretation process to the low-level picture processing stage.
That is, the system re-examines the segmented picture in detail
when it obtains some suggestions of errors from the knowledge sources
(object-detection sub-systems). As a result, some elementary regions come to be merged with neighboring regions or be split into
several small regions. (The detailed mechanism of this error correction will be described in Chapter 7.)

   As is obvious from the above description, our system bases its
interpretation of the scene on elementary regions. In the early
stages of image understanding research, edge-based interpretations
were often utilized to describe the structure of the scene. Brice
and Fennema [12] first made a region-based analysis of a picture.
In the case of such simple scenes as those of polyhedra, it is

both a natural and a good decision to use edge-based analysis of
the scene, since the knowledge about the scene itself is represented by the mutual relations among the edges of the objects to be
detected. In the case of complex natural scenes such as aerial
photographs, however, a region-based analysis is preferable because regions can represent various intrinsic properties of objects
such as shape, color, and texture. In the edge-based analysis, it
is very difficult to distinguish between boundaries of objects and
edges in textured regions.

This section describes the detailed algorithm of segmentation
in our system and shows the basic properties of the elementary regions calculated after segmentation.

4.2. Edge-Preserving Smoothing

A digital picture is corrupted by noise. On the other hand,
edges in a digital picture are blurred to some extent by the samp-
ling process. In order to partition a digital picture into regions
or to extract edges and lines, the noise and blur should be removed.
The ordinary smoothing by local averaging blurs sharp edges as well
as removes noise, so that we cannot always utilize it as the preprocessing for segmentation or edge extraction. Thus, we have devised a new smoothing method named "edge-preserving smoothing"
which not only removes noise in uniform areas but also sharpens
blurred edges between regions. (The details of this smoothing
method have been described in Section 3.1.)

Four digital pictures of an aerial photograph (in BLUE, GRFFN
RED and INFRARED hands), are smoothed by the edge-preserving smoothing method. Figure 4.1[1] shows the smoothed verion of the aerial
photograph shown in Fig. 2.1. Figures 4.2 (a) and (b) show cross-
sections of the picture in the BLUE band before and after smoothing, respectively, along the same horizontal line. Almost all
isolated noise is removed, and all edges are clearly sharpened,
which makes the subsequent segmentation process very reliable.

Of course, as the edge-preserving smoothing is in substance
an averaging over some local area, very thin lines and small regions are completely smoothed out. The minimum area size for a
region to survive after smoothing is about 3 × 3. Considering
that one point in a picture corresponds to 50 × 50cm on the ground,
objects smaller than 1.5 × 1.5m are too small to be detected from
a picture of this resolution. Thus, it might be safe to neglect
such small regions.

---

[1]For Fig. 4.1 see color insert

4.3   Segmentation

4.3.1.   Merging of Pixels and Labeling of Elementary Regions

As one can imagine from the cross section in Fig. 4.2(b), the smoothed picture in each spectral band is composed of a great number of small patches with constant gray level. The segmentation process in our system merges neighboring small patches with similar multispectral properties into one "elementary region", which becomes a basic unit for the subsequent analysis. This process does not use any knowledge about the scene to be analyzed, and relies only on the multispectral properties of the pixels in the smoothed picture. Each pixel in an elementary region is labeled with a unique region number, by which the analysis programs for feature extraction and object recognition identify the area of each elementary region in the two-dimensional space.

The algorithm for merging pixels and labeling regions is based on a simple region-growing method, and consists of the following steps:

Step 1:   If all pixels are labeled, then end. Else take an unlabeled pixel and assign a new unused region number.

Step 2:   If the differences of gray level in the four spectral bands between the new labeled pixel and its neighboring pixels (4-adjacency is used) are less than the thresholds θ_i (i = B, G, R, IR), respectively, then merge the neighboring pixels and assign them the same region number.

Step 3:   Iterate Step 2 until no pixels adjacent to the newly labeled region can be merged.

Step 4:   Go to Step 1.

As this region growing algorithm is very simple, it might seem to cause crucial merging errors. However, it can give a fairly good result without making serious errors. The reasons for this are:

(a) Blurred edges where incorrect merging is apt to take place have been clearly sharpened by the edge-preserving smoothing (see Fig. 4.2).

(b) The threshold value for each spectral band is adaptively determined from the picture data under analysis. (The algorithm for threshold determination will be described in the next section.) In addition,

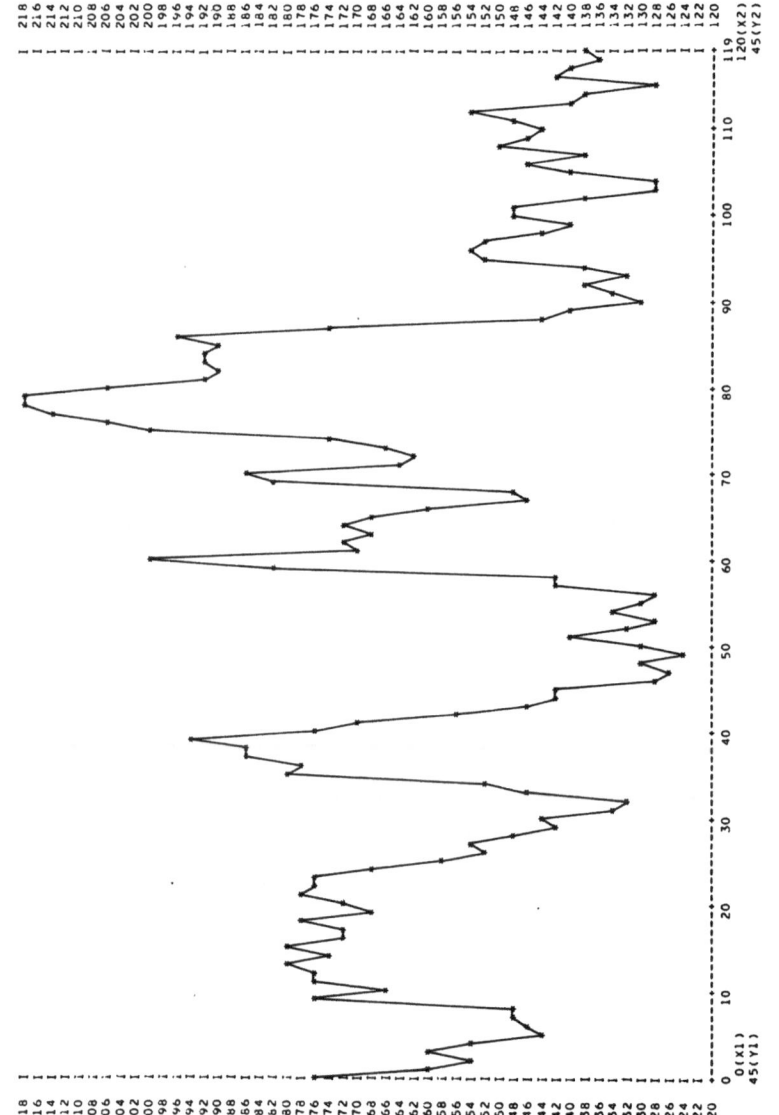

Fig. 4.2 (a) Cross section of the original picture in the BLUE band

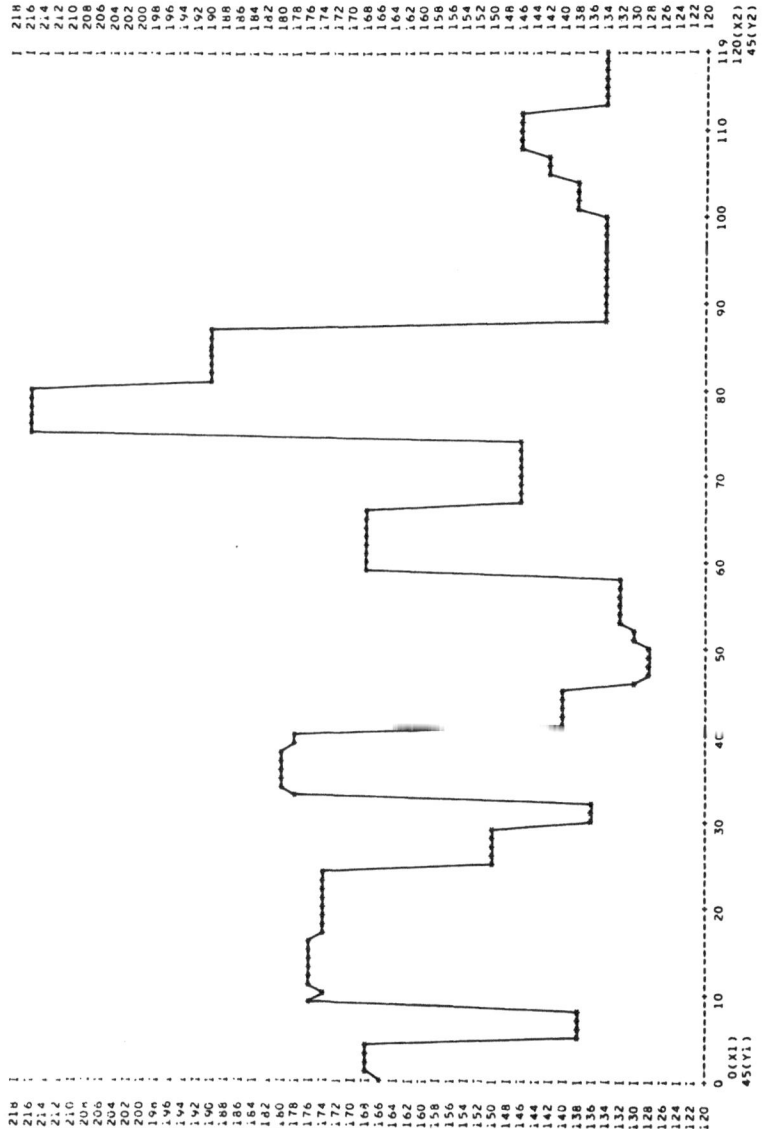

Fig. 4.2 (b) Cross section of the smoothed picture in the BLUE band; isolated noise is removed, and all edges are clearly sharpened

the threshold determination is made as moderate as
possible, since insufficient merging is preferable
to excessive merging for the later stages of analysis. It is much easier to merge small regions into
a "correct" region than to split a "false" region.

Figure 4.3[1] shows the result of merging pixels and labeling
regions. There exist many very small regions consisting of a few
pixels around the boundaries of larger regions. The reasons for
this are:

(a) As each picture in the four spectral bands is sampled
    independently of the others, it is almost impossible
    to avoid small registration errors between the four
    digital pictures. The accuracy of registration might
    be about one pixel.

(b) As discussed in Section 3.1, the edge-preserving
    smoothing sharpens blurred edges. As a result,
    pixels on a blurred edge come to be included into
    some homogeneous region along that edge (see Fig.
    3.7, where two of the three pixels on the blurred
    edge were included into the homogeneous region on
    the left side and the other into that on the right
    side). Since the smoothing is applied to the picture
    in each spectral band independently, the boundaries
    of the regions in the four pictures sometimes happen
    to deviate from each other.

These small displacements between the four pictures cause many
small regions in the segmented picture, because pixels to be merged
should have similar gray levels in all spectral bands. Considering
the sizes of the masks used in smoothing, small regions having less
than seven pixels can be regarded as being generated by these displacements; our smoothing method always smooths out regions smaller
than the mask size used for averaging, $i.e.$, seven pixels. Thus, we
merge these small regions with the neighboring large regions which
have the most similar multispectral properties.

Figure 4.4[1] shows the corrected result of Fig. 4.3. We call
this correctly segmented picture the "LABEL PICTURE" and regions
in the LABEL PICTURE "elementary regions". In this aerial photograph the LABEL PICTURE consists of 1,044 elementary regions, each
of which is labeled with a unique region number. Each characteristic region extracted and each object located in the subsequent

_____

[1]For Figs. 4.3 and 4.4 see color insert

analysis processes is represented as a set of these elementary regions. (Of course, sometimes elementary regions are modified by
the system in order to correct segmentation errors.)

4.3.2. Threshold Determination

The threshold value in each spectral band used for segmentation should be adaptively determined from the picture data under
analysis, because if the merging program used some predetermined
fixed value as a threshold, it would make serious mistakes in other
pictures.

Automatic threshold determination is one of the intensively
studied techniques in digital picture processing. There have been
proposed a variety of methods on this subject. (For a survey of
these methods see [82].) Most of the methods so far proposed deal
with pictures composed of two types of entities: object and background, and utilize a histogram of gray levels to obtain the threshold. It is usually determined as a valley between two peaks of the
histogram. Then this threshold is used to isolate regions with
gray levels smaller (larger) than the threshold as the objects.

As an aerial photograph contains a variety of objects with
different spectral properties, the simple valley detection in the
gray-level histogram, which is very popular for isolating objects
from the background, does not work well. Heavily textured regions
in a picture such as forest areas make it more difficult to find a
threshold automatically. We have devised a new method of threshold
determination to cope with such complex pictures as aerial photographs. It utilizes a histogram of differential values in the picture for each spectral band to find the threshold for that spectral
band.

The following shows the adaptive threshold-determination algorithm used in our system.

Step 1: Differentiate the smoothed picture of a spectral
band using the operator

$$d(i, j) = \max_{\substack{-1 \le k \le 1 \\ -1 \le \ell \le 1}} |x(i, j) - x(i+k, j+\ell)| , \qquad (1)$$

where $x(i, j)$ and $d(i, j)$ denote the gray levels
and the differential value at a point (i, j),
respectively.

Step 2: Divide the differentiated picture into sixteen
64×64 square blocks and make a histogram $h_n(d)$

of the differential values $d(i, j)$ in the n-th block $(n = 1, \ldots, 16)$.

Step 3:   THE VALLEY-DETECTION ALGORITHM

For each histogram $h_n(d)$, find the minimum value d_n which satisfies the following inequalities:

$$h_n(d_n) \le h_n(d_n + k) \quad \text{for all } k = 1, \ldots, N,$$

$$d_n > d_n^*, \tag{2}$$

where d_n^* denotes the differential value for which histogram $h_n(d)$ has the maximum population. N is set to nine in this experiment.

Step 4:   Find the minimum value among the d_n for all blocks $(n = 1, \ldots, 16)$ and make it the threshold value θ for all areas of the picture, that is

$$\theta = \min_{1 \le n \le 16} d_n. \tag{3}$$

We apply this threshold detection program to the picture in each spectral band; we then have the thresholds $\theta_i = (i = B, G, R, IR)$.

The basic idea of the valley-detection algorithm in Step 3 is as follows. The operator used for differentiation calculates the maximum difference (contrast) between a pixel and its neighbors. In homogeneous regions, it gives small values which come from noise fluctuations, while it gives large values around the boundaries between regions. As the number of pixels in homogeneous regions is much bigger than that around the boundaries, the histogram of differential values usually takes the form shown in Fig. 4.5. As the peak at the left side certainly comes from noise, we can safely set the threshold at the foot of this peak. We start from the peak of the histogram and move to the right while comparing the height with the heights on the right side of the histogram. At the descending slope of the peak, the height of the histogram is greater than almost all heights on the right side. Therefore, inequality (2) is not satisfied and we move one step to the right. On the other hand, at the foot of the peak, the height of the histogram is less than the heights on the right side, and inequality (2) is satisfied. Thus we can determine the foot of the peak in the histogram. This process is essentially equivalent to calculating the gradient of the histogram at each position and detecting the position where the

gradient is zero. This algorithm is insensitive to the irregular
peaks or valleys in the histogram because peaks or valleys whose
widths are less than N are neglected; the inequality (2) should be
satisfied for all k, $k = 1, \ldots, N$. Figure 4.6 shows the histogram
of the differential values in one of 16 blocks in the smoothed picture of the INFRARED band, where the value pointed to by the arrow
is selected as the threshold for this histogram.

The reasons why we divide the picture into 16 blocks and take
the minimum value among 16 local thresholds as the universal threshold value for a picture are:

(a) The histogram of differential values over the whole
    picture area does not have a sharp valley as in Fig.
    4.6 due to textured regions such as forest areas.
    Fig. 4.7 shows the histogram of the differential
    values over the whole picture area in the INFRARED
    band. It has a very gentle slope on the right side
    of the peak, which prevents us from finding the foot
    of the peak (threshold) in the histogram.

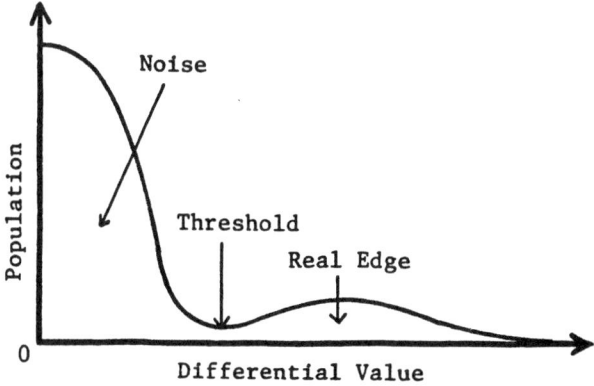

Fig. 4.5 The general form of the histogram of differential values

(b) The threshold value detected by this method is used
as the dissimilarity measure for the region-growing
process in segmentation. As the merging algorithm
used in our system has the tendency to merge pixels
over a region boundary, we should avoid this "mismerging" by selecting a conservative threshold.

We have analyzed several different pictures of aerial photographs (see Chapter 8). The results of the segmentation of these
pictures have shown that the series of processes for segmentation,
that is, edge-preserving smoothing, automatic threshold determination, and merging of pixels with similar multispectral properties,
works quite well in spite of variable qualities of pictures and
heavily textured regions. Although there exist some errors in
segmentation, they will be corrected at the high-level processing
stage using knowledge about objects.

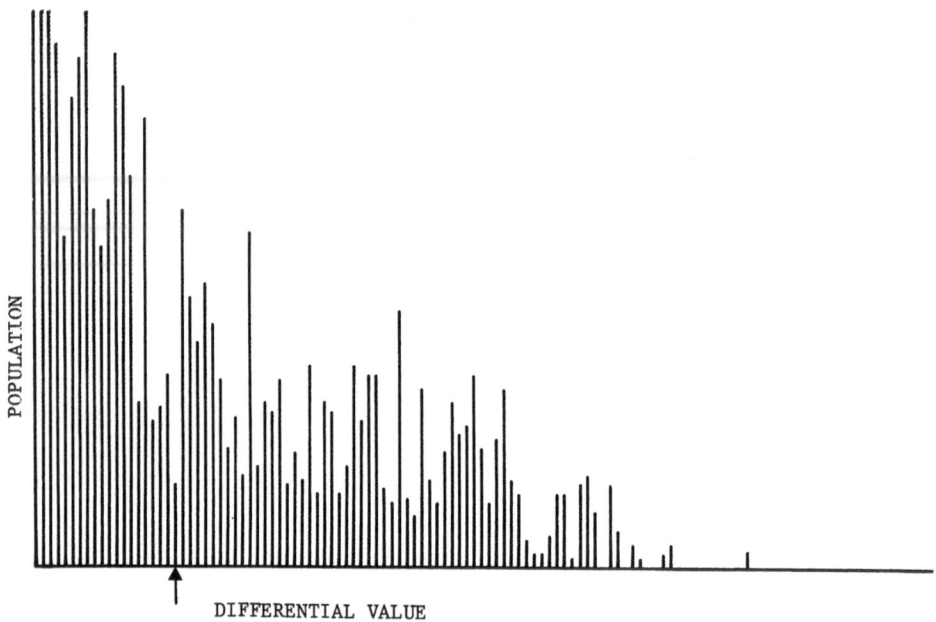

Fig. 4.6 Histogram of differential values in one of 16 blocks;
the valley detection algorithm selects 18 as the threshold for this histogram

4.4.   Calculation of Basic Properties of Elementary Regions

After segmentation, several basic properties of each elementary region, such as the average gray level in each spectral band,
area size, coordinates of centroid, location, and some fundamental
shape features are calculated and stored in "the property table"
in the blackboard (see Figure 2.2) together with the region number.
(The structure of the blackboard will be given in Section 7.1.)
These properties are very simple, and in order to recognize objects,
various specialized features of regions have to be calculated depending on the properties of the objects to be detected. Such specific features are to be calculated by object-detection subsystems
at the subsequent stages of the analysis.

The location of an elementary region in the LABEL PICTURE is
represented by two coordinate pairs, (BX, BY), (EX, EY), as shown
in Fig. 4.8. When one wants to get the two-dimensional image of
an elementary region, one needs only to scan within the rectangular

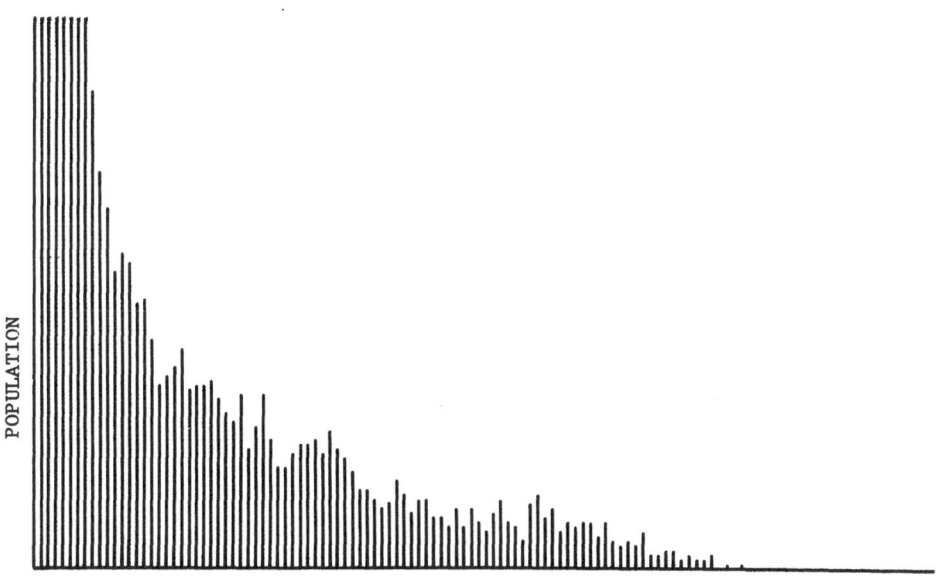

DIFFERENTIAL VALUE

Fig. 4.7 Histogram of differential values over the whole picture
          area; due to textured regions this histogram has no
          sharp valley

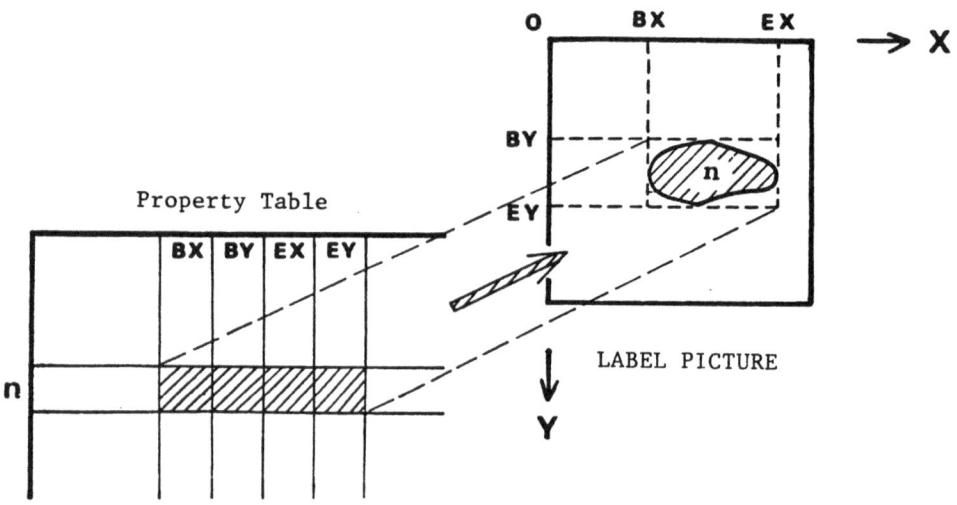

Fig. 4.8 Specification of the location of an elementary region in the LABEL PICTURE

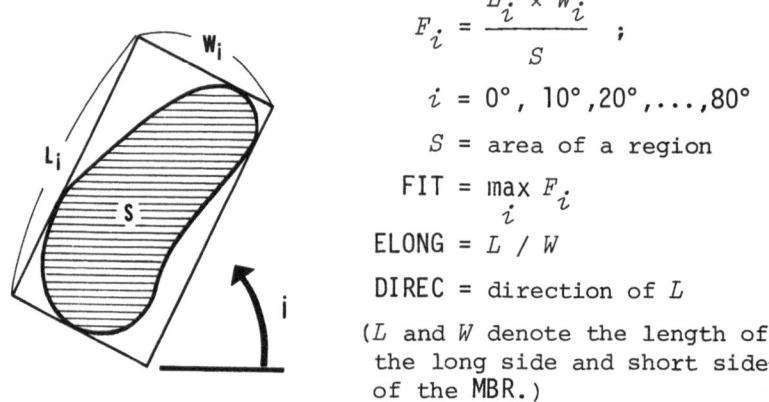

$$F_i = \frac{L_i \times W_i}{S} \quad ;$$

$i = 0°, 10°, 20°, \ldots, 80°$

S = area of a region

FIT = $\max_i F_i$

ELONG = L / W

DIREC = direction of L

(L and W denote the length of the long side and short side of the MBR.)

Fig. 4.9 Minimum bounding rectangle of a region

area specified by (BX, BY) and (EX, EY), which saves a great deal
of processing time.

The shape features of an elementary region calculated at this
step are FIT, ELONG, and DIREC. These features are calculated by
the following process:

(1) Calculate the minimum bounding rectangle (MBR) of a
region, where MBR denotes the minimum-area rectangle
which encases the region. MBR is defined as the one
whose F = (area of the region) / (area of the encas-
ing rectangle whose sides are parallel with the coordinate axes) takes the maximum when we rotate the
coordinates by $0°$, $10°$, ..., $90°$ (Fig. 4.9). We
define FIT as the maximum of F, which measures the
degree of matching the region to a rectangle.

(2) ELONG, the elongatedness of the region, is defined
by ELONG = L/W, where L and W denote the length of
the long and short sides of the MBR, respectively.

(3) DIREC denotes the direction of the long side of the
MBR.

All subsequent analysis programs take each elementary region
as an object or a part of object to be recognized. They consult
the property table in order to examine its properties or write in
the results of the analysis, and access the LABEL PICTURE to measure specialized shape features and to find the spatial relation-
ship among elementary regions.

5. EXTRACTION OF CHARACTERISTIC REGIONS

As discussed in Section 2.4, in the analysis of complex aerial photographs it is almost impossible to build an exact world model which can predict the locations of objects and guide the scene interpretation process as is done in the analysis of blocks-world scenes [69] and human faces [38]. In order to realize an efficient and reliable analysis, however, it is desirable to focus the analysis on some local area, where some object is assumed to exist, and to analyze it in detail using a special-purpose program designed for a specific object. Thus, we have introduced a focusing mechanism in the system for the structural analysis of aerial photographs.

At the first stage of the focusing process in our system (the global survey of the whole scene) we extract several kinds of "characteristic regions" by analyzing the segmented picture from various standpoints. These characteristic regions can be considered as useful clues for estimating approximate areas of objects. For example, a very elongated region, which is one of the characteristic regions, can be thought of as a "candidate region" for elongated objects such as roads, rivers, and railroads. At the detailed examination stage, object-detection subsystems, which perform the knowledge-based analysis, confine their sophisticated analysis within some local areas specified via various combinations of characteristic regions (see Fig. 2.5).

After segmentation, characteristic regions are extracted by a group of mutually independent characteristic-region extractors and the result of the analysis is stored in the blackboard, where each extracted characteristic region is represented as a set of elementary regions. The characteristic-region extractors do not use any knowledge about specific objects, but rely only on the general information about the properties of the picture data under analysis, such as the scale of an aerial photograph, the direction of the sun,

and the multispectral properties of some materials (not objects).
Some of this information comes from the photographic conditions
and the rest from physical experiments on spectral reflectances.

From the standpoint of the software structure of the system
each characteristic-region extractor is implemented as a separate
module which works independently of the others, just like the
object-detection subsystems (see Fig. 2.2). Thus, if needed, we
can easily modify any of them and add a new module to increase the
reliability of the process of estimating spatial domains of objects, which allows us to augment the system performance quite
easily.

Elementary regions formed by segmentation (the low-level processing) are regarded as the primitive symbols in terms of which
the various objects are described. After segmentation elementary
regions are characterized by a set of simple attributes (i.e., the
basic properties described in Section 4.4), but no organization
processes among them are performed at the segmentation stage. The
role of that stage is just to partition the raw picture data into
a set of elementary regions. The process of extracting characteristic regions can be regarded as that of structuring such symbols (elementary regions) and finding the rough structure in the
picture so as to facilitate the knowledge-based (high-level) processing. In this sense, this process may be named "the middle-
level processing" which, without using specialized knowledge of
objects, organizes the results of the low-level processing into
more useful forms for the high-level processing. Thus the raw
picture data is organized through two levels of processing before
the sophisticated knowledge-based processing is performed. As
these low- and middle-level processes do not rely on the specialized knowledge of objects, they will be applicable to any type of
scene. (Some of the characteristic-region extractors in the present system utilize the gray-level information in specific spectral bands, so that they will have to be modified when one wants
to analyze a picture taken through different spectral bands.)

In order to specify the spatial domains of objects as correctly as possible, we utilize such features as size, shape, brightness, multispectral properties, texture, and spatial relations
among the elementary regions. Based on these features, the present system extracts eight types of characteristic regions, namely,
large homogeneous regions, elongated regions, shadow regions,
shadow-making regions, water regions, vegetation regions, large-
vegetation areas, and high-contrast texture areas. Here we use
the word "area" when one characteristic region consists of many
elementary regions. That is, each of the first six types of characteristic region corresponds to a single elementary region while
both of the last two types consist of sets of elementary regions.
Table 5.1 summarizes the features used to extract these characteristic regions and the objects which are supposed to correspond

to each type of characteristic regions.

All the features used here are insensitive to the changes in photographic conditions, and all parameters used in processing are adaptively determined from the picture data under analysis. Therefore, the extracted characteristic regions can form a reliable basis for a subsequent detailed analysis.

In the following sections, we will describe the detailed algorithms for extracting these characteristic regions.

5.1. Large Homogeneous Regions

As shown in Fig. 4.4, aerial photographs are divided into a number of elementary regions at the segmentation stage. Since the land-use patterns in urban and suburban areas are very complex, most of the elementary regions become very small. In this situation, regions of large area size which show similar spectral properties extending very widely on the ground surface can be regarded as very outstanding characteristics. These large homogeneous regions in an aerial photograph may be considered as useful cues for recognizing large uniform objects such as crop fields, grasslands, lakes, and seas. (Fine textures of grasses and waves have been smoothed out during edge-preserving smoothing.)

When we are going to select large homogeneous regions from a group of elementary regions, we have to determine the threshold for area size. Since the result of partitioning the picture into elementary regions depends on the structure of the scene, we have no way of knowing *a priori* how large regions are present in the scene. Therefore, we cannot use a prespecified value as the threshold of area size. In order to make the system adaptive to various scenes, the threshold of area size should be adaptively determined from the picture under analysis.

The threshold value used to specify large homogeneous regions is determined by the following method:

Step 1: Make a histogram of the area sizes of the elementary regions.

Step 2: Apply the VALLEY-DETECTION ALGORITHM in Section 4.3 to this histogram.

Step 3: Use the detected area size as the threshold for selecting large homogeneous regions.

The histogram of the area sizes of the elementary regions usually takes the form shown in Fig. 5.1, where the large peak on the left comes from many small regions. The VALLEY-DETECTION ALGORITHM starts from the top of this peak and finds the location of its foot.

Table 5.1 Characteristic regions; the present system extracts eight types of characteristic regions based on such features as size, shape, brightness, multispectral properties, texture, and spatial relationships among elementary regions.

PROPERTY	CHARACTERISTIC REGION	OBJECTS SUPPOSED TO BE INCLUDED
SIZE	Large Homogeneous Region	Crop Field, Sea, Lake
SHAPE	Elongated Region	Road, River Railroad
BRIGHTNESS	Shadow Region	(River)
LOCATION	Shadow-Making Region	Building, House, Tree
	Vegetation Region	Crop Field, Grassland, Forest
SPECTRAL INFORMATION	Large Vegetation Area	Crop Field, Grassland, Forest
	Water Region	Sea, Lake, River
TEXTURE	High Contrast Texture Area	Forest, Residential Area

As the threshold of area size is adaptively determined from the picture data, we extract as the large homogeneous regions those elementary regions which are relatively larger than the others. Thus the altitude from which an aerial photograph is taken is immaterial. Figure 5.2 shows the large homogeneous regions extracted from the picture in Fig. 4.4. There are 14 regions which correspond to crop fields, bare soils, grasslands, and roads (see the original aerial photograph in Fig. 2.1).

5.2. Elongated Regions

There exist many elongated objects such as roads, rivers, and railroads in aerial photographs. These objects tend to be segmented into several elongated regions because of small objects on them *e.g.* cars, bridges, and shadows. Such shape features as elongatedness are very insensitive to changeable photographic conditions and can be regarded as useful clues to locate objects. Therefore, we extract the elongated regions as "candidate regions" for the recognition of elongated objects.

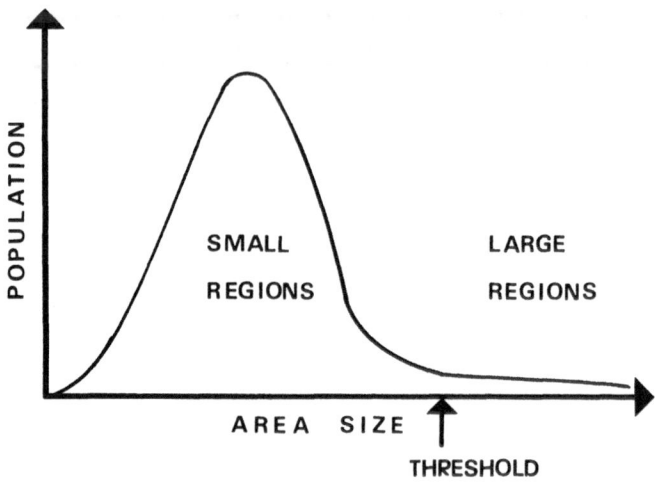

Fig. 5.1 The general form of the histogram of area sizes

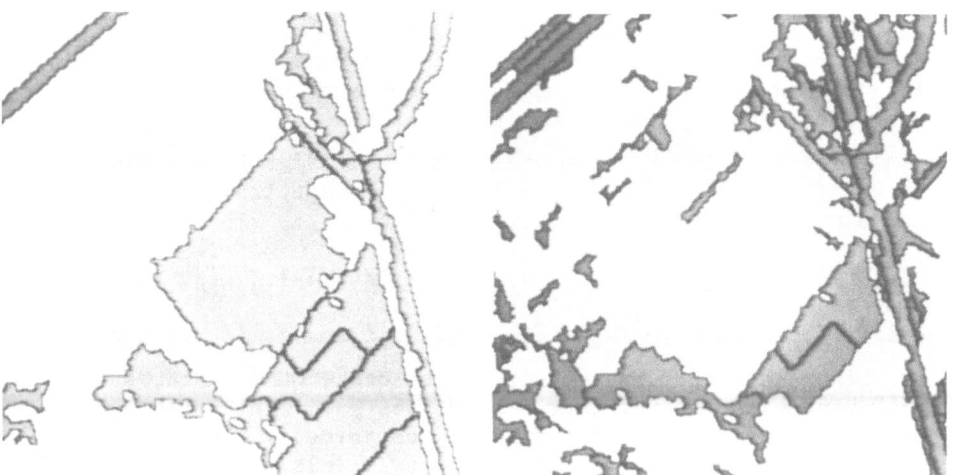

Fig. 5.2 Large homogeneous   Fig. 5.3 Elongated regions
          regions

First, elementary regions whose ELONGs (see Section 4.4) are
greater than 3.0 are extracted as elongated regions. However, as
ELONG measures the elongatedness based on the MBR of a region, it
does not represent the correct elongatedness for curved regions.
That is, ELONG has meaning only if the degree of matching a region
to a rectangle, FIT, is large. Thus, for those regions whose FIT
is less than 0.5, we calculate a new elongatedness, ELONG 2, based
on the longest path on the skeleton of the region. (See Section
3.2.)

Fig. 5.3 shows the extracted elongated regions whose ELONG
or ELONG 2 is greater than 3.0. We can see that almost all regions
corresponding to roads are successfully extracted.

5.3. Shadow Regions and Shadow-Making Regions

Most of the objects in aerial photographs can be considered
as two-dimensional objects, e.g. crop fields, grasslands, seas,
and so on. In order to recognize houses, buildings, and trees,
however, it is desirable to know that the regions representing
these objects have three-dimensional structures, that is, they have
heights. However, since the heights of these objects are very
short compared to the altitude from which an aerial photograph is
taken, the depth in the scene becomes very small. And we have no
means to measure the depth. We have to look for some other cues

to know that an object has height. We utilized the shadow as a clue for getting the three-dimensional information about objects. That is, first we extract shadow regions and then estimate the corresponding "shadow-making" regions, which represent objects with heights, by using their spatial relationships with the shadow regions.

In order to extract shadow regions, we first make a picture of brightness from smoothed pictures in the four spectral bands. Brightness of a point (i, j) is usually defined by

$$I(i, j) = \frac{1}{4} (X_B(i, j) + X_G(i, j) + X_R(i, j) + X_{IR}(i, j)) \qquad (1)$$

where $I(i, j)$, $X_B(i, j)$, $X_G(i, j)$ $X_R(i, j)$, and $X_{IR}(i, j)$ denote the brightness and the gray levels in the BLUE, GREEN, RED, and INFRARED bands respectively, at a point (i, j). Since the light of longer wave length is not apt to be scattered by the atmosphere below an airplane, the differences in gray levels between shadow regions and others is especially prominent in the RED and INFRARED bands. Therefore, in order to enhance this difference, we define the brightness by

$$I(i, j) = \frac{1}{6} (X_B(i, j) + X_G(i, j) + 2X_R(i, j) + 2X_{IR}(i, j)). \qquad (2)$$

Fig. 5.4(a) and (b) show the pictures of the brightness values calculated by equations (1) and (2), respectively. One can see that the shadow regions in Fig. 5.4(b) are very clearly enhanced.

The threshold of brightness is also adaptively determined by the method described below.

Step 1:   Make a histogram of the brightness I of the smoothed pictures.

Step 2:   Calculate the average brightness I in the whole area of the picture.

Step 3:   Calculate the brightness I_1 which makes the between-class variance maximum when the histogram between the lowest brightness and I is divided into two classes [55].

This method is formally described as follows: Let the histogram have the L discrete levels $[1, 2, \ldots, L]$ and let n_i denote the population at level i. The between-class variance $\sigma_B^1(k)$ is calculated by the following equation when we divide the histogram into two classes at level k, that is, $[1, 2, \ldots, k]$ and $[k+1, \ldots, L]$:

Fig. 2.1 A picture of an aerial photograph

Fig. 4.1 Smoothed version of the picture of Fig. 2.1

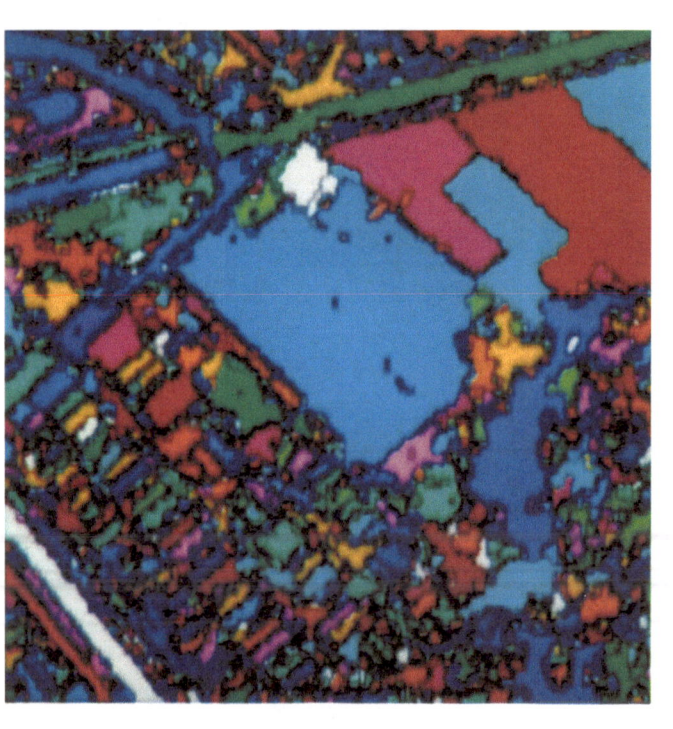

Fig. 4.3   Result of merging pixels and labeling
regions; there exist many small regions
around the boundaries of larger regions. These small regions result from
displacements among the four digital
pictures (color has no significance)

Fig. 4.4 (a) The LABEL PICTURE; small regions in
Fig. 4.3 are merged into neighboring
larger regions with similar multispectral properties. This picture consists
of 1,044 "elementary regions," each of
which is labeled with a unique region
number

(b) Result of segmentation; the original picture is overlaid with the boundaries of elementary regions

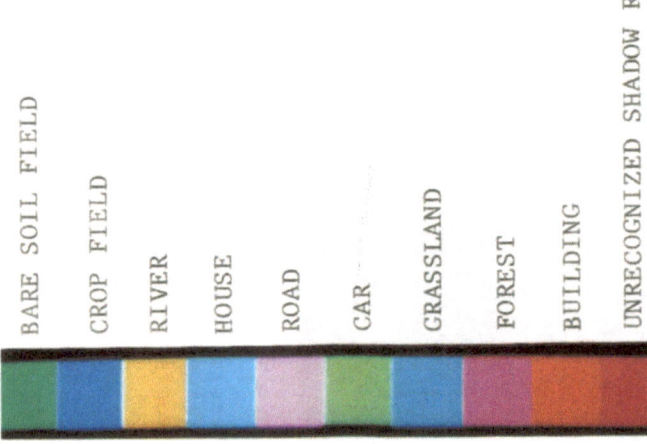

BARE SOIL FIELD
CROP FIELD
RIVER
HOUSE
ROAD
CAR
GRASSLAND
FOREST
BUILDING
UNRECOGNIZED SHADOW REGION

Fig. 8.2 Correspondence between the object categories and the colors

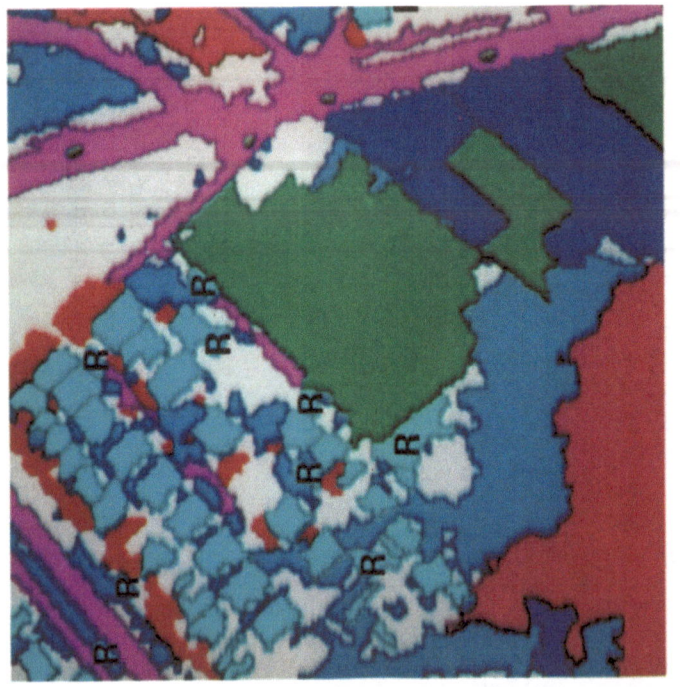

Fig. 8.1 Final result of analyzing the picture shown in Fig. 2.1; "R" denotes misrecognized houses

(a)

(b)

Fig. 8.3 Some examples of aerial photographs of urban and suburban districts;
(a) C4-8 (b) C4-7 (c) C3-6 (d) C1-2

(d)

(c)

Figure 8.3 (continued)

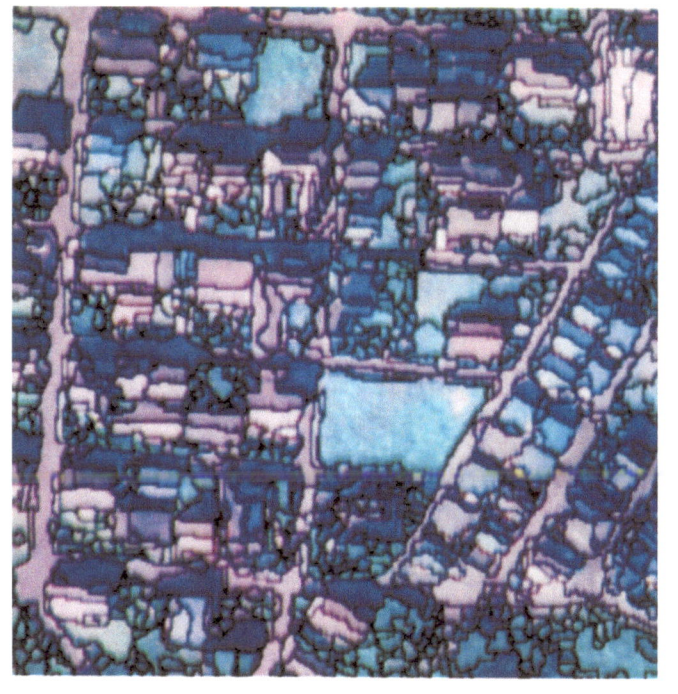

(b)

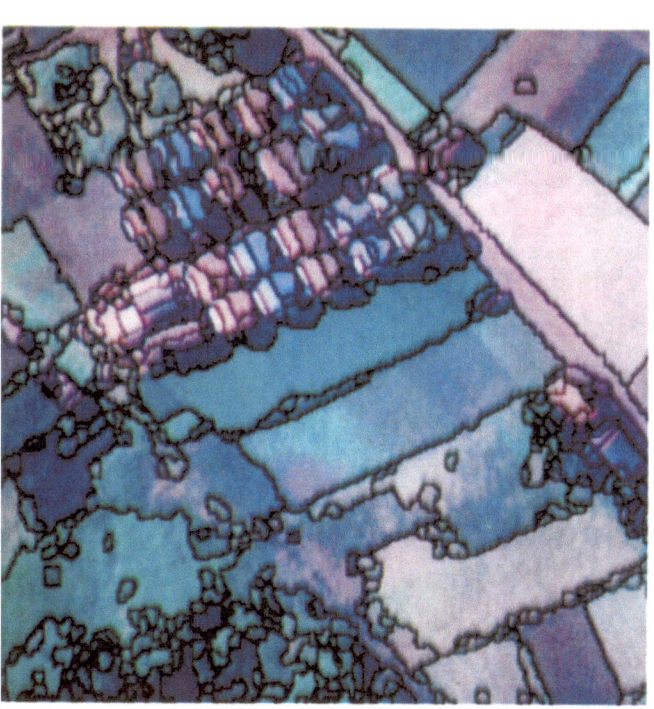

(a)

Fig. 8.4 Results of segmentation of the pictures in Fig. 8.3

(d)

(c)

Figure 8.4 (continued)

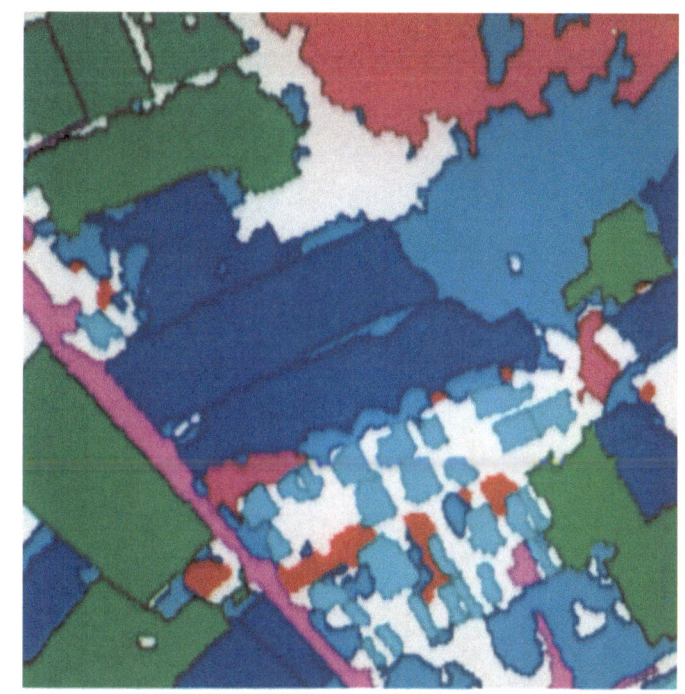

(a)

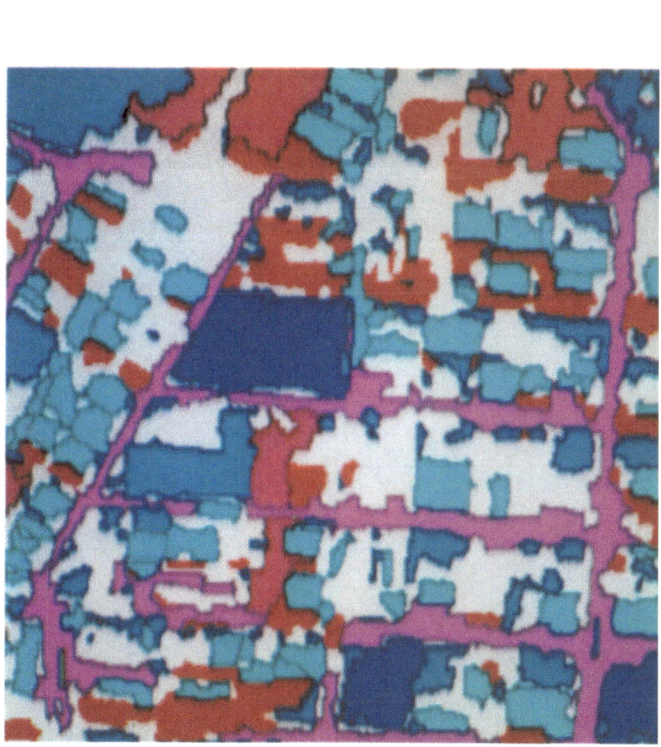

(b)

Fig. 8.5 Results of the analysis of the pictures in Fig. 8.3

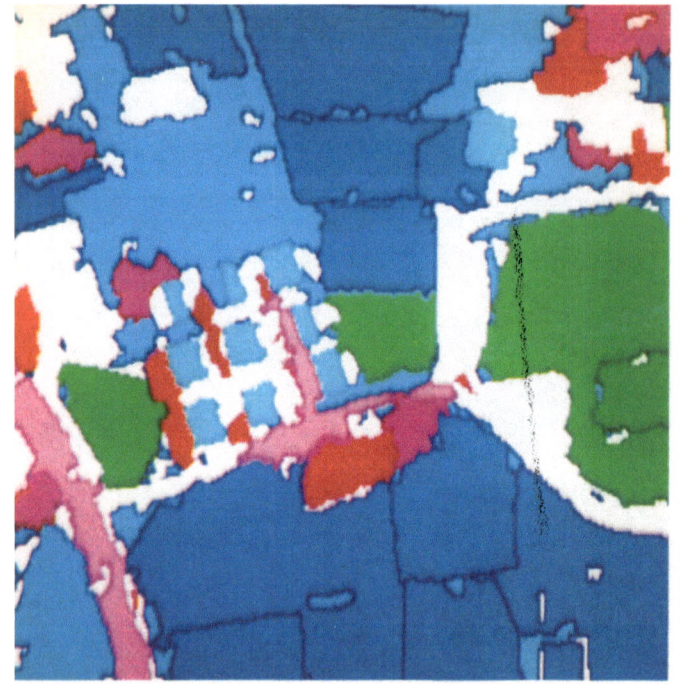

(d)

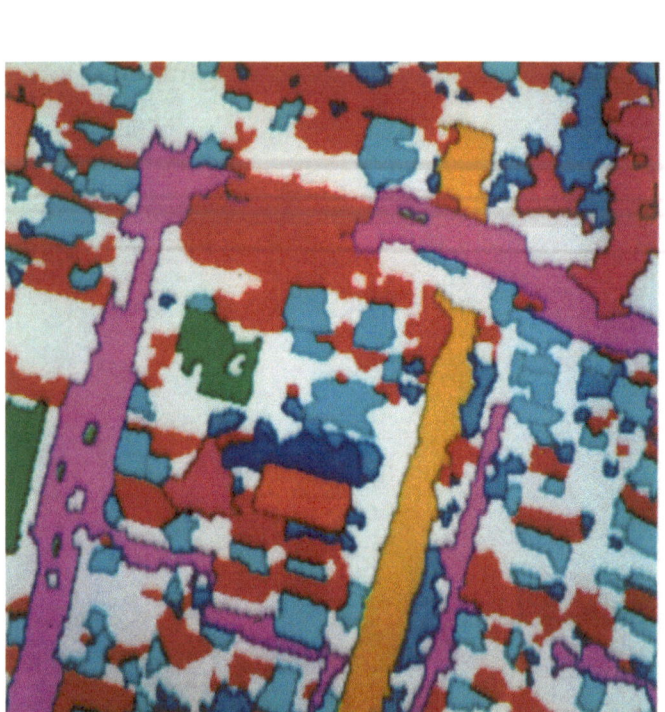

(c)

Figure 8.5 (continued)

(a)   (b)

Fig. 5.4   (a) Picture of brightness values calculated by equation
            (1)

           (b) Picture of brightness values calculated by equation
            (2); shadow regions are clearly enhanced

$$\sigma_B^2(k) = \frac{[\mu_T w(k) - \mu(k)]^2}{w(k)[1 - w(k)]} \quad ,$$

where
$$w(k) = \sum_{i=1}^{k} P_i \quad ,$$

$$\mu(k) = \sum_{i=1}^{k} i\, P_i \quad ,$$

and
$$\mu_T = \mu(L) \quad ,$$

where
$$P_i = \frac{n_i}{\sum_{i=1}^{L} n_i}$$

Then the optimal threshold k^* is determined as the level which makes $\sigma_B^2(k)$ maximum:

$$\sigma_B^2(k^*) = \max_{1 \le k \le L} \sigma_B^2(k)$$

Step 4: Calculate the gradient of the histogram at I_1.

Step 5: If the gradient is almost 0, I_1 becomes the threshold brightness I^*. Else search the valley of the histogram around I_1 by the VALLEY-DETECTION ALGORITHM in Section 4.3, and set the brightness of the valley as the threshold I^*.

Step 6: The elementary regions whose average brightness values are less than I^* are regarded as shadow regions.

Figure 5.5 and Fig. 5.6 show the brightness histogram and the shadow regions extracted by this method, respectively.

Next we extract shadow-making regions by using spatial relationships with the shadow regions, where we suppose that there are no shadow regions due to the clouds in the sky, that is, all shadow regions result from the three-dimensional objects on the ground surface.

Step 1: Extract the regions which are adjacent to shadow regions in the direction of the sun.

Step 2: Select the regions from those extracted in Step 1 which have a long common boundary with a neighboring shadow region in the direction away from the sun.

We regard the regions extracted in Step 2 as shadow-making regions. Figure 5.7 shows the shadow-making regions (enclosed by black lines) and corresponding shadow regions (gray shaded). They nicely correspond to trees in the forest area and houses in the residential area (see Fig. 2.1).

These shadow and shadow-making regions play an important role in the discrimination between three-dimensional objects and two-dimensional flat objects. The direction of the sun used in this process is one of the parameters concerning the photographic conditions, which are given *a priori*.

5.4. Vegetation Regions

The multispectral camera on an aircraft records the light

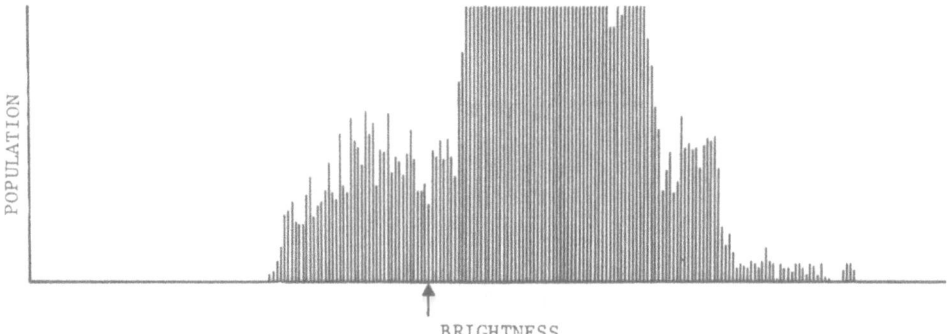

Fig. 5.5 The brightness histogram; the brightness indicated by the
arrow is selected as the threshold for extracting shadow
regions

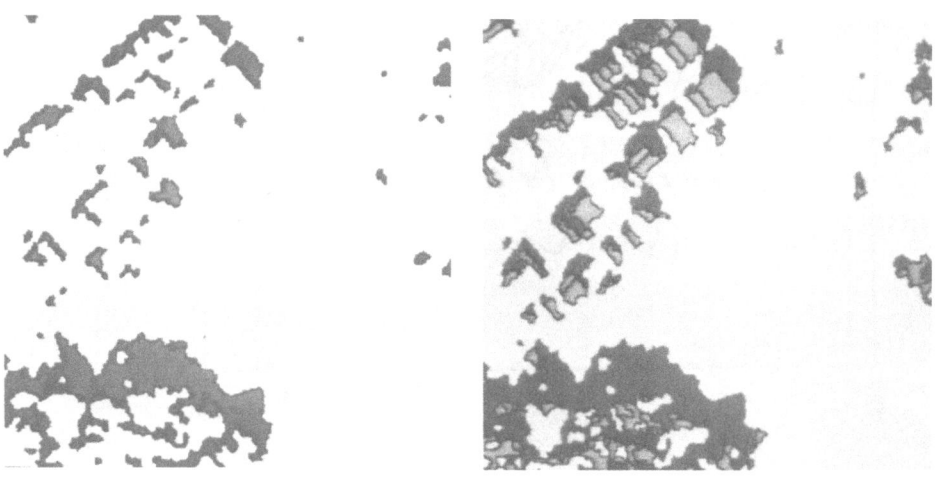

Fig. 5.6 Shadow regions   Fig. 5.7 Shadow-making regions; the
                                        regions enclosed by black
                                        lines show the shadow-
                                        making regions and those
                                        shaded gray denote the
                                        shadow regions

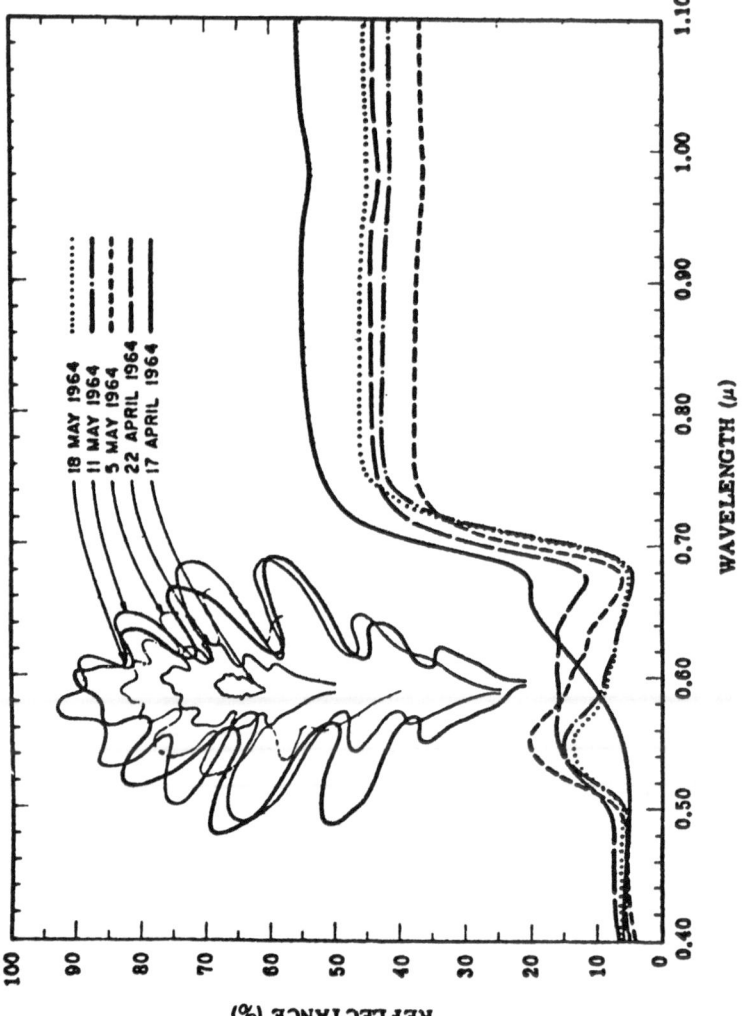

Fig. 5.8 General spectral characteristics of vegetation (from [25])

energy in multiple spectral bands which is radiated from objects
on the ground surface. Usually, the differences in the materials
of the objects are fairly well reflected by the multispectral
aerial photographs. Therefore, we want to use some characteristics
in the multispectral properties which are useful for recognition
of objects. But as the light energy recorded by the camera originally comes from the sun, various factors such as the season, time,
and weather can affect the observed data. In addition, as the
light passes through the atmosphere at least twice, the amount of
the energy recorded by the camera is heavily dependent on the path
length and the atmospheric conditions. Therefore, an object on the
ground surface does not always show the same multispectral characteristics. Thus we have to find materials which show prominent
and stable characteristics in multispectral properties without
being affected by such factors.

After several experiments on the multispectral properties of
various materials, we have devised two sets of features which stably
characterize vegetation and water areas, respectively, This section describes the method used to extract vegetable regions, and
the next section will be devoted to water regions.

Vegetation areas show very prominent characteristics in the
multispectral properties, so that we can easily extract vegetative
regions. Figure 5.8 shows the general spectral characteristics of
the vegetation (from [25]), from which we can see that the reflectance rate in the RED band is much smaller than that in the INFRA-
RED band, and that this characteristic is not affected by the
season.

We use $X_{IR}(i)/X_R(i)$ as the measure of the likelihood of vegetation, where $X_{IR}(i)$ and $X_R(i)$ denote the average gray levels of
an elementary region in the INFRARED and RED band, respectively.
Though one could conceivably use the difference, i.e., $X_{IR}(i)$ -
$X_R(i)$, as a measure, the difference between gray levels is very
sensitive to photographic conditions and shadows. Figure 5.9 (a)
and (b) denote the spectral properties of vegetations in sun and
in shadow, respectively, measured from the picture shown in Fig.
2.1. From these graphs we can see that the ratio between the
INFRARED and RED bands is very useful for extracting vegetation
regions even if they are in shadow. We determined the threshold
of the ratio to be 1.2 so that we can extract even very sparsely
vegetated areas. This threshold has been proved to be very stable
and effective for extracting vegetation from various aerial photographs.

Several experiments showed that blue roofs of houses have the
same characteristics as vegetation in the RED and INFRARED bands.
Figure 5.10 shows the multispectral properties of the blue roofs

in Fig. 2.1. As the color of the roof is very bright blue, the gray level in the RED band is lowered, which results in raising the ratio $X_{IR}(i)/X_R(i)$. Therefore, in order to correctly extract vegetation regions, we exclude those regions which have very high gray levels in the BLUE band, because vegetation regions do not have very high values in that spectral band. The threshold in the BLUE band is determined in exactly the same way as the extraction of shadow regions, except that the histogram of the gray levels in the BLUE band is ordered from high to low.

Figure 5.11 shows the vegetation regions extracted by this method. The vegetation regions are successfully extracted even if they are in the shadow, and no other erroneous regions are extracted.

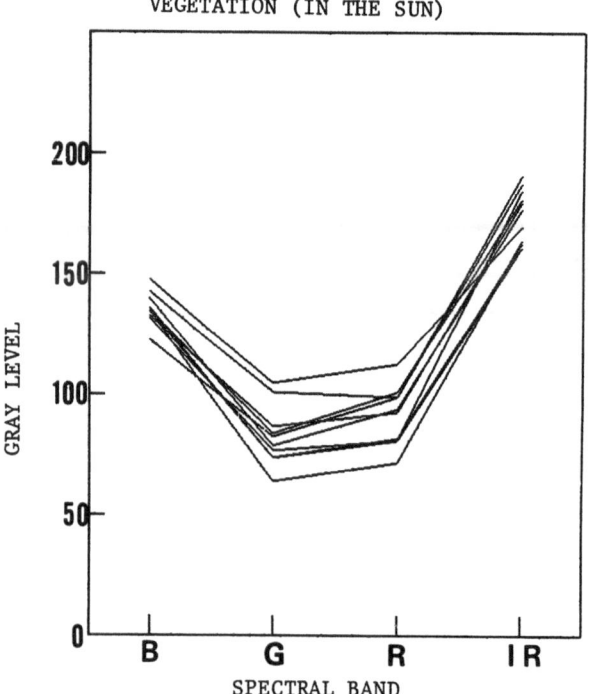

Fig. 5.9   (a) Spectral characteristics of vegetation in sun

After extracting individual vegetation regions, adjacent vegetation regions are merged into one region. If the areas of the
merged regions are greater than the threshold used in the extraction of large homogeneous regions, they are registered as large
vegetation areas, which will be used to specify the locations of
large planted areas such as crop fields, grasslands, and forest
areas. Figure 5.12 shows the large vegetation areas extracted
from Fig. 5.11. Each of the large vegetation areas consists of
many elementary regions.

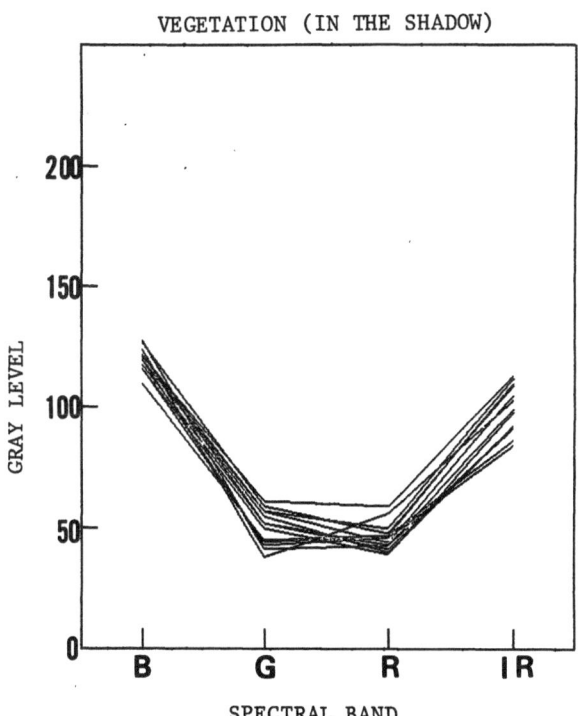

(b) Spectral characteristics of vegetation in shadow;
    the spectral characteristics of vegetation are pro-
    minent even in shadow

5.5. Water Regions

Even though the picture shown in Fig. 2.1 does not contain water areas, it is very important to extract water regions to discriminate between rivers and roads, and sea and ground.

It has been shown through various experiments on the spectral reflectance of water that the amount of light energy reflected by pure water decreases gradually as the wave length of the light becomes longer, and that in the INFRARED band most of the light energy is absorbed by water. This characteristic is also true for sea water. Figure 5.13(a) shows the multispectral properties of water regions measured in the picture shown in Fig. 8.3(c), which show good agreement with the general spectral characteristics of water.

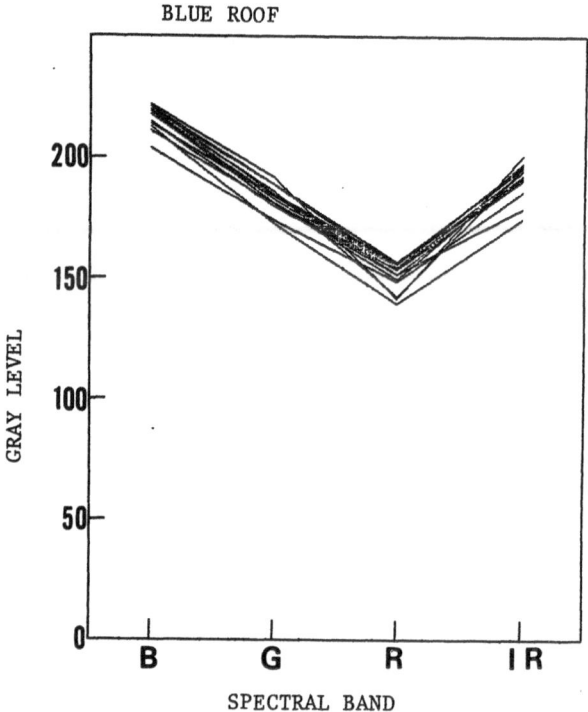

Fig. 5.10 Spectral characteristics of blue roofs; the ratio between the INFRARED and RED bands is large, just like vegetation

From the above observations, we regard those regions as water regions which satisfy the following conditions:

1. The brightness of a region is lower than the average brightness of the whole picture.

2. The average gray levels in the four spectral bands satisfy the following three inequalities simultaneously:

$$1.3[X_{IR}(i)] < X_R(i) < 1.9[X_{IR}(i)]$$

$$1.5[X_{IR}(i)] < X_G(i) < 2.0[X_{IR}(i)]$$

$$2.0[X_{IR}(i)] < X_B(i)$$

Figure 5.14 shows the water regions extracted from Fig. 8.3 (c). (No water regions have been extracted from the picture shown in Fig. 2.1.) While the water regions in the sun have been successfully extracted, those in the shadow have failed to be extracted, since the multispectral properties of water are greatly affected by shadow. (See Fig. 5.13(b), which shows the multispectral properties of water regions in shadow.) Moreover, if there are many weeds in the water, the gray level in the INFRARED band increases, and, as a result, the extraction of water regions will fail. In this regard, the features used here are too strict for extracting water regions under various conditions. If the conditions are relaxed, however, some erroneous regions will be extracted

Fig. 5.11 Vegetation regions   Fig. 5.12 Large vegetation areas; each area consists of many elementary regions

as water regions. Since the processes of object recognition rely
heavily on the characteristic regions, completely reliable water
regions must be extracted in this process. False water regions
should not be extracted because the object-detection subsystems
have no way of correcting such errors. The water regions left unextracted at this stage will be analyzed by using the information
about their environments, such as the spatial relationships with
the water regions extracted at the high-level processing stage.

5.6. High-Contrast Texture Areas

There are many small objects in the forest and residential
areas such as trees, houses, roads, and shadows. As a result,
these areas show great variety in color and brightness, and become

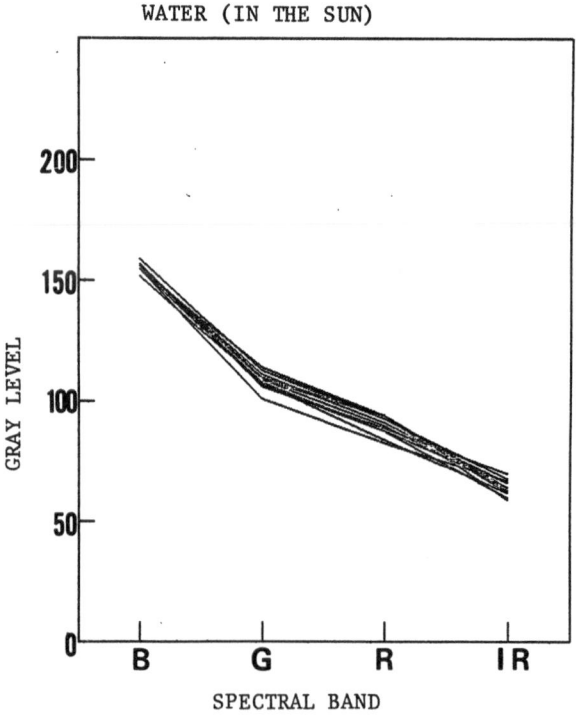

Fig. 5.13 (a) Spectral characteristics of water regions (in sun)

divided into many small regions by segmentation. It is almost impossible to try to give object labels to these small regions by
merely using their own properties. For example, as a forest area
has a heavy texture, it is segmented into a number of small regions.
No matter how intensively we examine the properties of each small
region, such as shape, size, and color, we cannot recognize the
forest area unless we extract an aggregate of the small regions.
Therefore, first we extract a spatial set of small regions which
corresponds to forest and residential areas as a whole, and then
go into the recognition of each constituent region based on the
properties of the group of small regions. We call an area in which
many small regions are clustered a high-contrast texture area.
(Fine textures have been smoothed out by edge-preserving smoothing,
so that the textured areas extracted here are those with high-
contrast coarse textures, such as forest and residential areas.)

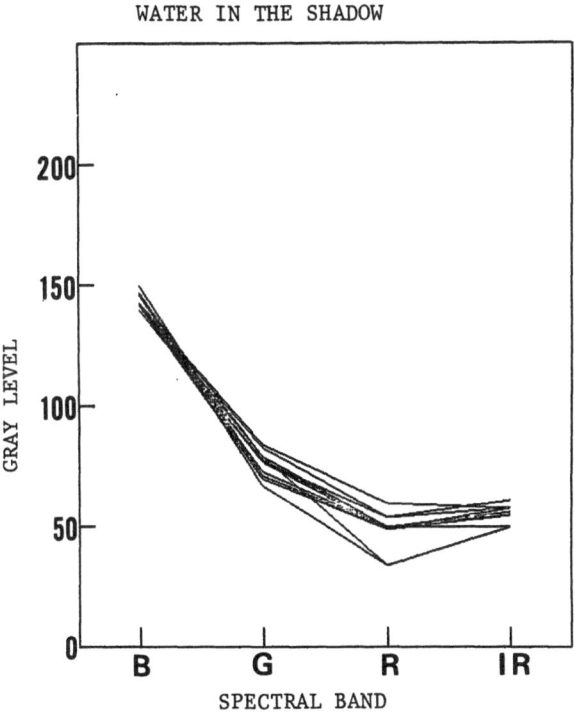

(b) Spectral characteristics of water regions in shadow;
    the spectral characteristics of water are greatly
    affected by shadow

Fig. 5.14 Water regions extracted
from the picture in
Fig. 8.3(c); the water
regions in shadow are
not extracted

We use the density of the boundaries of elementary regions to
extract high-contrast texture areas.

Step 1:   Extract the boundaries of elementary regions
          (Fig. 5.15(a)).

Step 2:   Move an $N \times N$ window over the picture of region
          boundaries, and if the window contains more than
          $2N$ boundary points, mark the central point of
          the window (Fig. 5.15(b)). N is set to 15 in
          this picture, which corresponds to 7.5m on the
          ground surface.

Step 3:   Enlarge (two steps) those points extracted at
          Step 2, then shrink (four steps) and re-enlarge
          (two steps). By this enlarge-shrink-enlarge
          operation, small holes and thin peninsulas
          whose widths are less than four are removed
          (Fig. 5.15(c)).

Step 4:   Extract elementary regions more than half of
          whose areas are included in the areas extracted
          at Step 3, and merge adjacent elementary regions
          into one area.

Step 5:   Register the merged areas as high-contrast tex-
          ture those areas whose sizes are greater than
          the threshold of the area size used in the ex-
          traction of large homogeneous regions. (Fig.
          5.15(d)).

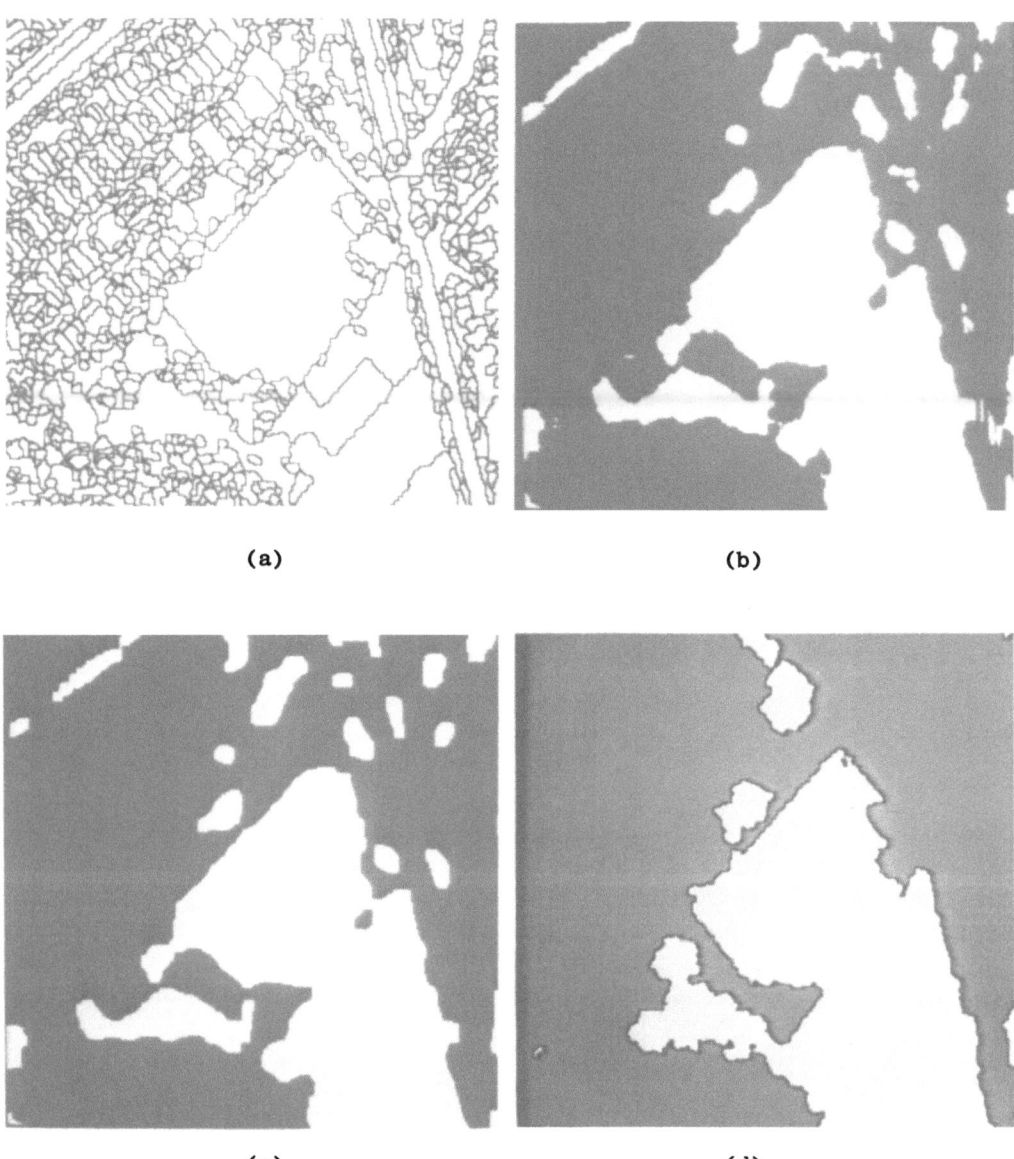

(a)   (b)

(c)   (d)

Fig. 5.15   (a) Boundaries of elementary regions

(b) The areas with high density of boundary points

(c) Result of removing small holes and thin peninsulas

(d) High-contrast texture area; this area consists of
    elementary regions

The purpose of Step 4 is to adjust the boundary of a high-contrast texture area so as to make it to coincide with those of elementary regions, and to represent each high-contrast texture area as a set of elementary regions.

The high-contrast texture area shown in Fig. 5.15(d) contains both forest areas and residential areas (see Fig. 2.1), which will be discriminated by combining the high-contrast texture area with other types of characteristic regions.

6. OBJECT RECOGNITION

As there are many objects of various kinds on the surface of the ground, diverse knowledge should be incorporated to describe its structure. As mentioned in Section 2.3, the software architecture of a production system has the desirable characteristics enabling the analysis of such a complex scene as an aerial photograph. The diverse knowledge, which can hardly be represented in a compact form, is divided into a set of mutually independent knowledge sources (production rules), in which we can write individually the specialized knowledge of each specific object. A knowledge source for an object is in our system an object-detection subsystem. It checks the contents of the blackboard and returns the result of the analysis to the blackboard. It never calls other subsystems directly. All communication among object-detection subsystems is made indirectly via the blackboard.

After the extraction of characteristic regions, a group of object-detection subsystems is put to work to recognize objects of various kinds (see Fig. 2.2). Each subsystem is designed to locate objects of a specific kind, using the specialized knowledge about their intrinsic properties and the environments in which they are embedded.

Each object-detection subsystem consists of three processing stages: the condition check, the calculation of new specialized features, and the recognition judgement. At the first stage of the analysis it checks various features of regions already calculated, to find the local areas where specific objects are highly likely to exist. Then it applies the specialized feature-extraction programs to the selected regions for making a judgement on whether or not they are the objects to be located (see Fig. 2.5). It is very time-consuming and often wasteful to calculate the specialized features required to recognize specific objects for all regions in the picture. Therefore each object-detection subsystem selects at the condition check stage the "candidate regions" for the subsequent specialized analysis. If there are no regions which satisfy the conditions, the later parts of the object-detection subsystem, *i.e.*, the specialized feature calculation stage and the recognition judgement stage, are not activated.

The types of object-detection subsystems can be divided into two categories according to the information they use in selecting the candidate regions for objects to be detected.

TYPE 1. Picture data-driven subsystems

The subsystems of this type check the existence of the local areas with specified properties by combining several kinds of characteristic regions. That is, the candidate regions for objects are specified in terms of logical combinations of characteristic regions. For example,

candidate regions for crop field = (large homogeneous region) ∧

(vegetation region) ∧ ($\overline{\text{water region}}$) ∧ ($\overline{\text{shadow-making region}}$)

where ∧ and ‾‾‾‾‾‾‾ denote "and" and "negation", respectively. Of course, in order to recognize a candidate region as a crop field, some specialized features should also be calculated before proceeding to the recognition judgement stage.

TYPE 2. Model driven subsystem

As mentioned before, in order to analyze complex aerial photographs, it is necessary to utilize the contextual information of the environment besides the knowledge about the intrinsic properties of objects. The subsystems of this type pick up the candidate regions by using the spectral and spatial relationships with the objects which have been already recognized by other subsystems, and then go into the recognition process considering the environmental information as well as the intrinsic properties of objects.

With the help of model-driven subsystems, the system can successfully locate the context-sensitive objects such as cars, and can perform a "heterarchical" analysis to raise the reliability and efficiency of object detection. That is, object-detection subsystems of type 1 usually use very strict conditions for recognizing objects so as to avoid errors, *i.e.*, the recognition of false objects. Consequently, the efficiency of recognition is lowered, and some objects are left unrecognized. Then model-driven subsystems analyze these "ambiguous" objects by consulting the properties of the already recognized objects. Thus we can realize a very efficient object recognition without decreasing the reliability of the analysis.

In the analysis of aerial photographs, we often encounter cases where a new method needs to be devised to find an object or new kinds of objects appear in a scene. As each object-detection subsystem is a module which works independently of the others, it is very easy to modify its content or to add a new object-detection subsystem. The modification of the subsystems does not cause any unexpected side-effects. The modularity of the system at the object recognition stage is crucial because we have to make repeated trial and error experiments in order to examine what knowledge and which properties are useful for locating various kinds of objects.

The present system has sixteen object-detection subsystems for nine kinds of objects: crop field, bare soil field (crop field without plants), forest, grassland, road, river, car, building, and house. Table 6.1 summarizes the types of characteristic regions used by each object-detection subsystem for selecting candidate regions for objects.

In the following sections, we will describe the detailed algorithms for these object-detection subsystems.

6.1. Crop Field

A crop field is considered as a flat region having a large area, a compact shape, and a straight region boundary because it is an artificial object. The subsystem for the recognition of a crop field, CF1, first picks up those regions which satisfy the following logical expression as the candidate regions for the crop field.

(large homogeneous region) ∧ (vegetation region) ∧

$\overline{\text{(water region)}}$ ∧ $\overline{\text{(shadow-making region)}}$.

The other properties not specified here are considered as "don't care" conditions. For example, a candidate region of a crop field may or may not be an elongated region. If a region touches the picture frame and its area size is greater than half of the threshold area for large homogeneous regions, it is also included as a candidate region (of course, the other properties, *i.e.*, (vegetation region) $\overline{\text{(water region)}}$ ∧ $\overline{\text{(shadow-making)}}$ should be satisfied). This is because we want to recognize a crop field even if it happens to be cut off by the picture frame.

Then CF1 checks the compactness and the straightness of the

boundary of each candidate region by the following method.

Step 1: If the measure of noncompactness of the candidate region, B^2/S, is greater than 35.0, reject
the region, where B and S denote the length of
the boundary and the area of the region. (By
this criterion, irregular-shaped regions and
very elongated regions, which show large values
for noncompactness, are removed from the candidate regions for the crop field.)

Note: A very popular measure for the noncompactness of a
region is (the length of a region boundary)$^2/$
(the area size). In the Euclidian space it takes
the minimum value of 4π when the region is a
circle. (Rosenfeld [63] has discussed some problems which occur in digital pictures.) While it
is not an information-preserving shape feature,
that is, the original boundary cannot be restored
from it, it is invariant under any linear transformation (e.g. scaling, translation, and rotation).
Thus it is quite widely used as a shape feature to
discriminate silhouettes of objects.

The threshold used here, i.e., 35.0, is determined
experimentally.

Step 2: For each point P_i ($i = 1, \ldots, N$) on the boundary
of the region, calculate the angle Δ_i between
two straight lines connecting P_i with P_{i-n} and
P_{i+n} (Fig. 6.1), where $i-n$ and $i+n$ are calculated modulo N. Here n is set to 5, which corresponds to 2.5m on the ground.

Step 3: Let $\hat{N}$ denote the number of points where $|\Delta_i| \leq$
$22.5°$ ($|\cdot|$ denotes the absolute value). If $\hat{N}/N$
≥ 0.6, then the region is regarded as having a
straight boundary.

If these conditions are satisfied, the candidate region is
recognized as a crop field and the result is written in the blackboard. But if the shape of the region is unsuitable for a crop
field (while the other properties are satisfactory), the object-
detection subsystem returns the recognition status as "irregular-
shaped" to the blackboard. This status denotes that the region is
not recognized solely because of its irregular shape. Then the
system activates a split/merge program to modify the initial

Table 6.1 The types of characteristic regions used by the object-
detection subsystems

○ MUST BE   Note: "not vegetation region"
X MUST NOT BE   implies "not large vegeta-
— DON'T CARE   tion area" as well

Object Category	Object-Detection Subsystem	Large Homogeneous Region	Elongated Region	Shadow Region	Shadow-Making Region	Vegetation Region	Large Vegetation Area	Water Region	High-Contrast Texture Area
Crop Field	CF1	○	—	—	x	○	—	x	—
	CF2	—	—	—	x	○	—	x	—
Bare-soil Field (Crop Field without plants)	BS1	○	—	x	x	x	—	x	—
	BS2	—	—	x	x	x	—	x	—
Forest	FOREST	x	—	—	—	—	○	x	○
Grassland	GRASS	—	—	—	x	○	—	x	—
Road	RD1	—	○	—	—	x	—	x	—
	RD2	—	○	—	—	x	—	x	—
River	RV1	—	○	—	—	x	—	○	—
	RV2	—	—	—	—	x	—	○	—
Car	CAR	x	—	x	—	x	—	x	—
Building	BUILDING	—	—	—	○	x	—	x	—
House	RESIDENTIAL AREA	x	—	—	—	—	x	—	○
	HOUSE1	—	—	x	○	x	—	x	—
	HOUSE2	—	x	x	○	x	—	x	—
	HOUSE3	—	—	x	—	x	—	x	—
	HOUSE4	—	x	x	—	x	—	x	—

segmentation assuming that the irregular-shaped region is caused
by a segmentation error. (The detailed process of correcting segmentation errors will be described in Section 7.3.)

BS1, the subsystem for a bare-soil field (here we use "bare-
soil field" to denote not a wasteland but a crop field without
plants) performs the same analysis processes as CF1 except that the
candidate regions for the bare-soil field are neither vegetation
regions nor shadow regions.

CF2 and BS2 belong to the model-driven subsystems and utilize
the knowledge that crop fields and bare-soil fields are often adjacent to each other. That is, if a region is adjacent to a crop
field or bare-soil field which have already been recognized, then
the region can be a candidate for a crop field or a bare-soil field
even if it is not a large homogeneous region. (But the area should
be greater than half of the threshold area for large homogeneous
regions.)

Figure 6.2 (a) and (b) shows the crop fields and the bare-soil
fields located by CF1 and BS1, respectively. (In this picture, no
crop field or bare-soil field has been newly recognized by CF2 and
BS2.)

6.2. Forest and Grassland

As a forest area is composed of a group of trees, it is considered to show heavy texture and the multispectral properties of vegetation. Thus, we can specify the candidate "areas" for forests
by the following combination of characteristic regions:

(high contrast texture area) $\wedge$ (large vegetation area) $\wedge$

$\overline{\text{(large homogeneous region)}}$ $\wedge$ $\overline{\text{(water region)}}$.

Figure 6.3(a) shows the areas characterized by the above conditions.
Since a forest can be considered to occupy a large area, we extract
large connected areas from Fig. 6.3(a). We select those areas whose
area sizes are greater than half of the threshold area for large
homogeneous regions. Note that each candidate area for the forest
consists of a set of elementary regions (Fig. 6.3(b)).

As is obvious from Fig. 6.3(b), the extracted candidate areas
include both forest areas and grasslands (see Fig. 2.1). In order
to discriminate trees from grasses and extract only forest areas,
we utilize shadow-making regions.

Step 1: Remove those candidate areas which do not contain any shadow-making regions.

Step 2: Extract the mixed areas consisting of shadow and
shadow-making regions from the candidate areas
and recognize them as forest areas.

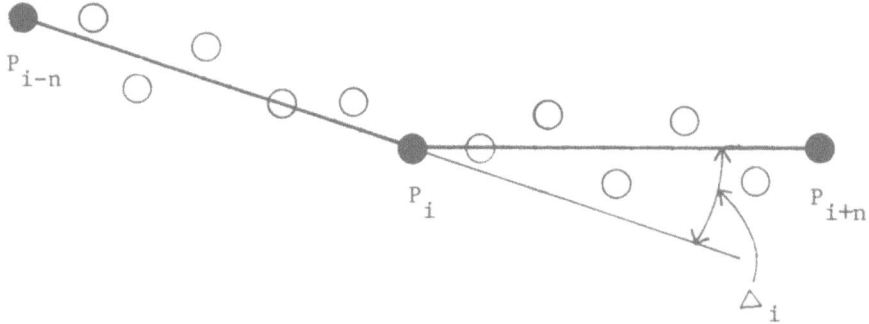

Straightness measure = $\hat{N}$ / N

$\hat{N}$; the number of points on the boundary where $|\Delta_i| \le 22.5°$

N ; the total number of points on the boundary

Fig. 6.1   Calculation of the straightness of a region boundary
(see text)

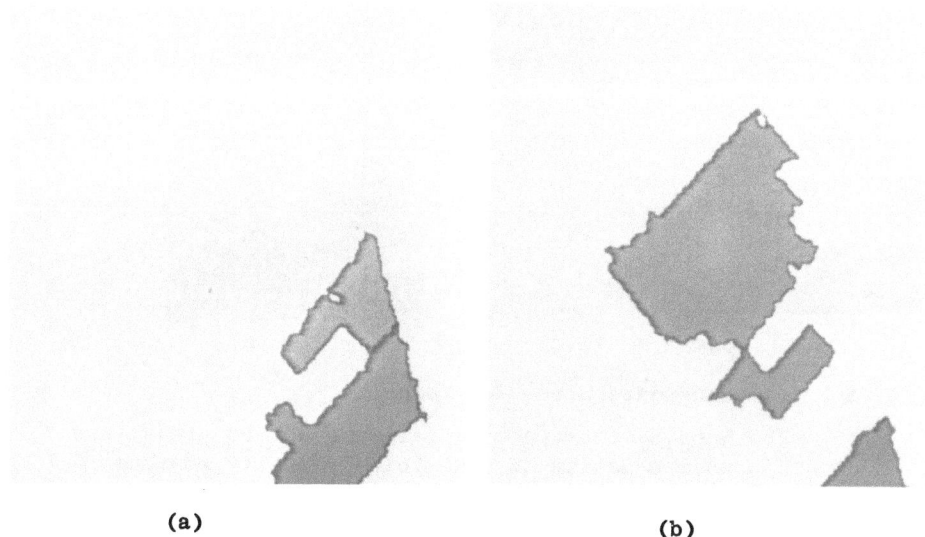

(a)   (b)

Fig. 6.2   (a) Recognized crop fields

(b) Recognized bare soil fields

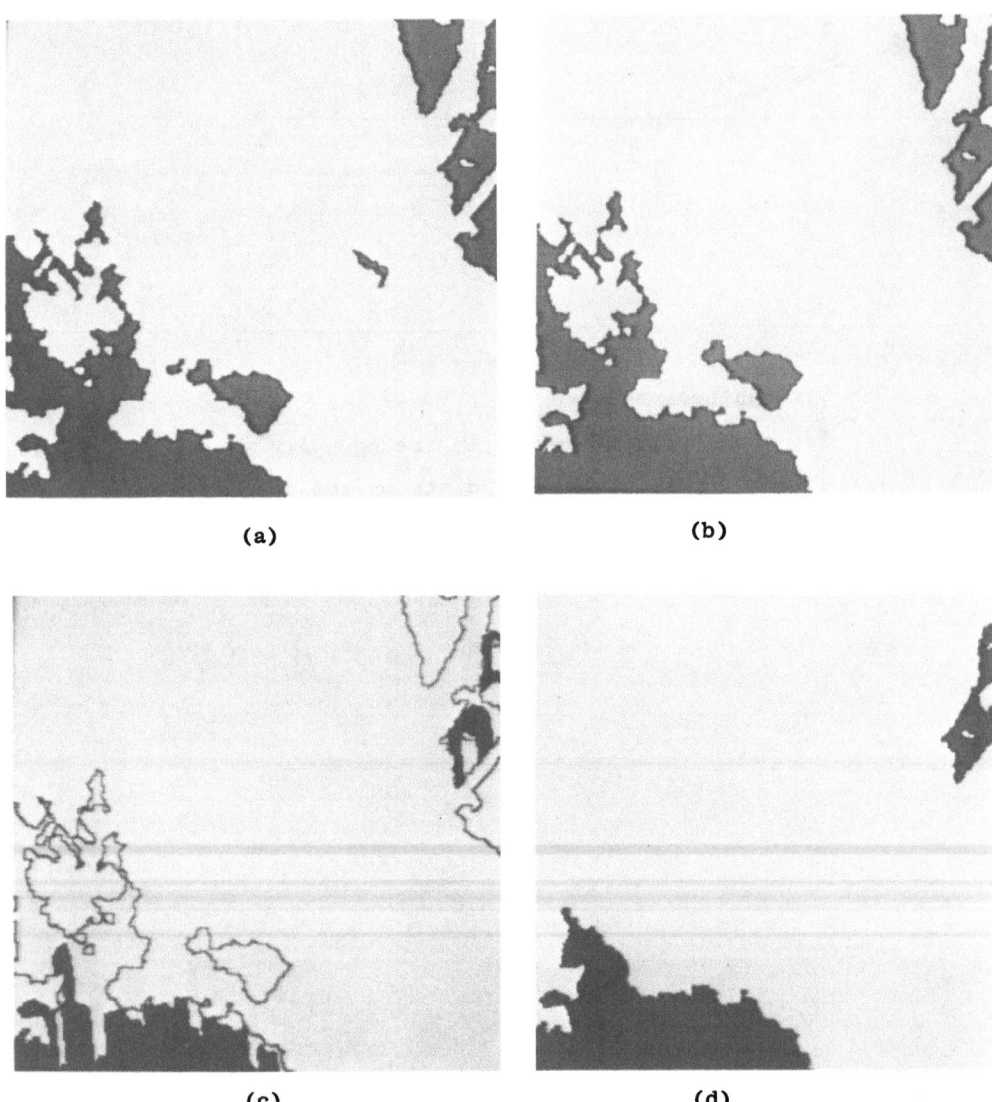

(a)

(b)

(c)

(d)

Fig. 6.3 (a) Candidate areas for forests

(b) Result of extracting large connected areas; each
    area consists of a set of elementary regions

(c) Result of extending shadow-making regions in candi-
    date areas

(d) Recognized forest areas

(i) Extend shadow-making regions contained in the candidate area in the direction of the sun until they
come across the boundaries of the candidate area
(Fig. 6.3(c)).

(ii) Apply the enlarge-shrink operation to the areas extracted by (i) and adjust the boundaries of these
areas with those of elementary regions. (This operation is the same as that used for extracting high-
contrast texture areas.) Then recognize large areas
among those extracted as forest areas (Fig. 6.3(d)).

By this process we can successfully distinguish forest areas from
adjacent grasslands.

Each recognized forest area consists of a number of elementary
regions (compare Fig. 6.3(d) with Fig. 4.4). In this case, an object which represents the group of elementary regions is generated
in the blackboard, and the object and its constituent elementary
regions are connected by "part-whole" relations. (For the detailed
structure of the blackboard, see Section 7.1.)

GRASS, the subsystem for grassland, picks up candidate regions
which are characterized by

(vegetation region) $\wedge$ $\overline{\text{(water region)}}$ $\wedge$ $\overline{\text{(shadow-making region)}}$.

After extracting candidate regions, GRASS merges neighboring
candidate regions into one grassland. The recognized grasslands
are also represented as groups of elementary regions. Figure 6.4
shows the recognized grasslands.

Comparing Fig. 6.4 with Fig. 6.2(a) and Fig. 6.3(d), we can
see that some elementary regions have been recognized as grassland
as well as crop field and forest area. This happens because each
object-detection subsystem recognizes objects without regard to the
results by the other subsystems. These contradictions (the multiple recognition of an elementary region) will be resolved by the
conflict-resolution mechanism of the system. (This mechanism will
be described in Section 7.2.) In the final result of the analysis
all recognized regions are labeled with a unique object name.

6.3. Road and River

It is very important to analyze rivers, roads, and railroads
in an aerial photograph. Their spatial characteristics such as
shapes and spatial relationships are quite stable over a long
period, and characterize the global structure of the scenes. Using
such information, we can connect neighboring frames of aerial photographs and can register pictures which are taken on different
days.

Fig. 6.4
Recognized grasslands; there
are some regions which are recognized as crop field and
forest area as well as grassland. These conflicts are resolved by the system

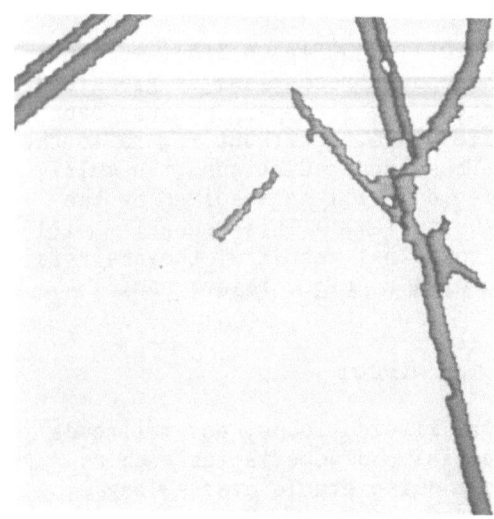

Fig. 6.5
Roads recognized by the RD1
subsystem; each road corresponds to one elementary region

RD1, the subsystem for road recognition, extracts the candidate regions for roads which satisfy

(elongated region) ∧ (vegetation region) ∧ (water region)

Then, it checks the elongatedness, ELONG2, and the variance of the
widths of the region along the longest path on the skeleton (see
Section 3.2) for each candidate region. RD1 recognizes the region
as a road if ELONG2 ≥ 8.0 and the variance of the widths is very
small (that is, its width is constant).

Since the area shown in the picture used here is rather small
(the altitude from which the aerial photograph has been taken is
very low), it often happens that roads are cut off by the edges of
the picture frame. Thus, RD1 also recognizes a candidate region
whose elongatedness is greater than 6.0 as a road if either end of
the region touches the picture frame. Figure 6.5 shows the roads
recognized by the RD1 subsystem. In this case each road corresponds
to an elementary region.

A road is often divided into several elongated regions in the
process of segmentation because of cars and shadows on it. Since
usually these elongated regions are not very long (ELONG2 ≥ 8.0),
RD1 does not recognize them as roads. Therefore, we have to connect
these regions into one region to recognize it as a road. The RD2
subsystem connects a pair of candidate regions for the road which
satisfy the following conditions:

(1) The differences of the average gray levels in the four
    spectral bands are smaller than θ_i (i = B, G, R, IR),

    respectively. (θ_i's denote the threshold values used

    in segmentation.)

(2) The ratio between the widths of the two regions is
    between 2/3 and 3/2.

(3) The shortest distance between the end points of the
    longest paths on the skeletons of the two regions
    is less than $3W$, where W denotes the smaller width
    of the two regions.

(4) The directions of the regions coincide with each
    other.

If the above conditions are satisfied, the pair of candidate
regions are connected into one region, and the gap between them is
filled by the following method:

Step 1:   Connect by a straight line two points on the
          longest paths of the two regions which are lo-
          cated (width)/2 away from the end points (Fig.
          6.6(a)). (The reason we do not connect the

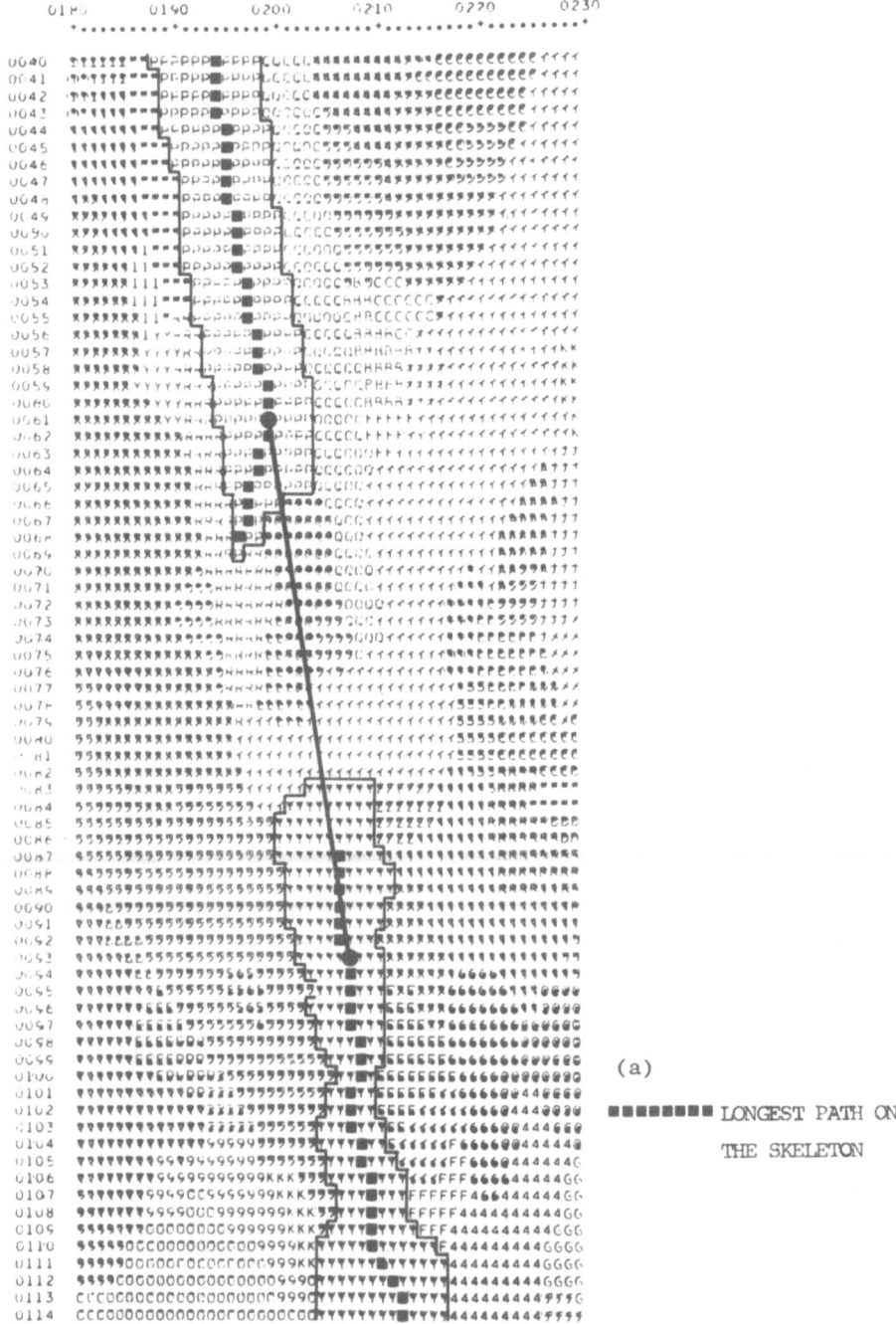

(a)

■■■■■■■■ LONGEST PATH ON
THE SKELETON

Fig. 6.6 Filling the gap between two candidate regions for roads
          (see text)

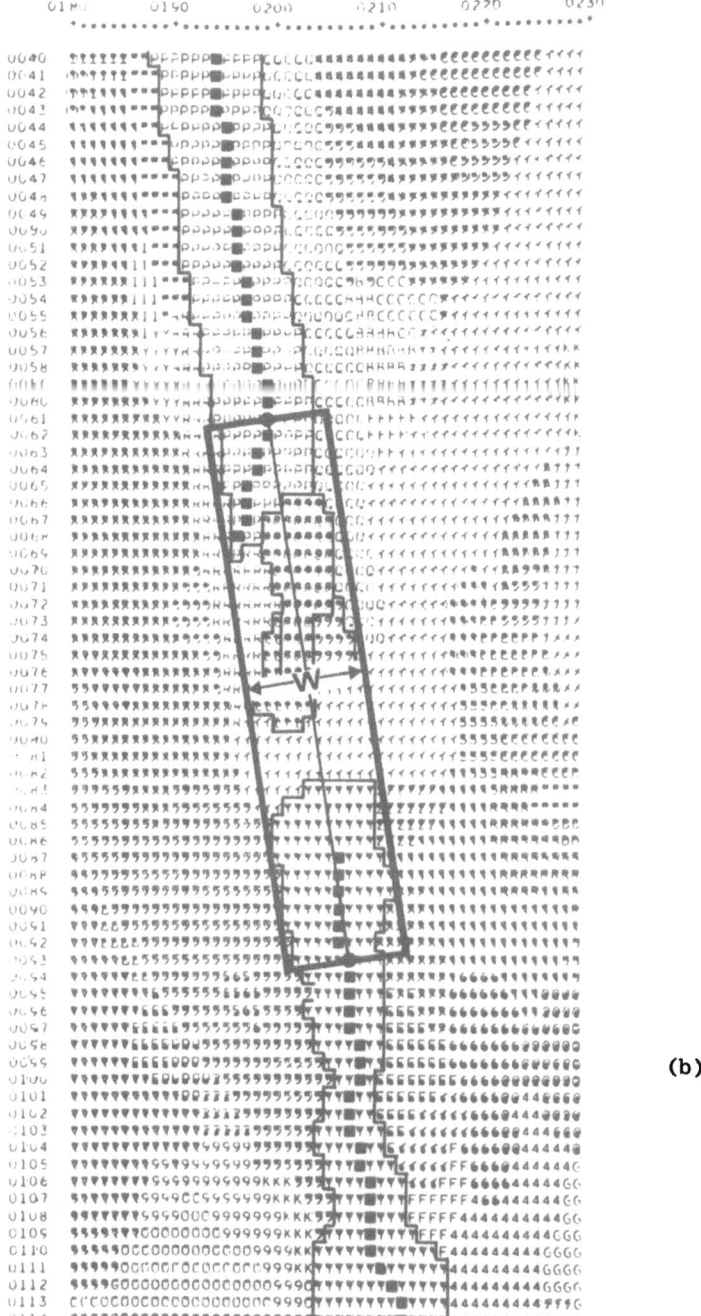

Fig. 6.6 contd.

(b)

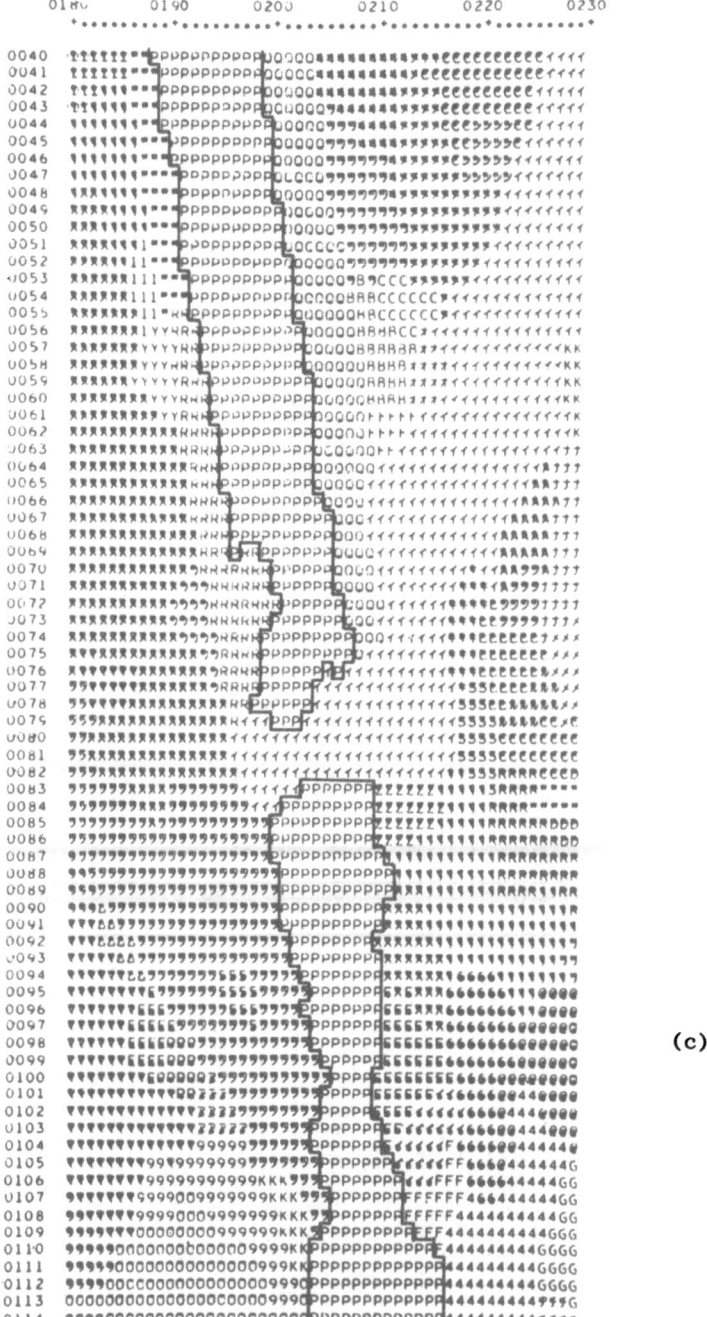

Fig. 6.6 contd.

> end points of the longest paths is that the
> longest paths often deviate from the medial
> axes near the edges of the regions.)

Step 2: Expand the straight line by W (Fig. 6.6(b)).

Step 3: Elementary regions which fall more than 50%
into the area generated by Step 2 are considered
as parts of the road if they are neither vegetation nor water regions (Fig. 6.6(c)).

After connecting candidate regions for roads, the RD2 subsystem
measures the elongatedness and width of the connected region. If
the elongatedness exceeds 8.0 and the width is constant, the connected region is recognized as a road. Thus the road connected by
this method consists of several elementary regions. In the two-
dimensional space, however, these elementary regions are not always
connected because of the conditions in Step 3, that is, the elementary regions located at the gap between two candidate regions may
be vegetation or water regions, or may not fall more than 50% into
the "gap area" generated by Step 2. (See Fig. 6.6(c).)

By applying the whole process described above to every pair
of candidate regions, all collinear "road segments" are connected
into a road. Then the RD2 subsystem picks up an unrecognized candidate region which has the following relationships with an already
recognized road, and connects it with the road as a side road.

(1) The difference in hue is small.

Let B_i, G_i, and R_i $(i=1,2)$ denote the average gray
levels of regions 1 and 2 in the BLUE, GREEN, and
RED bands respectively. We consider that the regions
1 and 2 have similar hue if the following inequalities are satisfied for each spectral band.

$$\frac{X_1 - \Theta_X}{T_1 + \Theta_X} \leq \frac{X_2}{T_2} \leq \frac{X_1 + \Theta_X}{T_1 - \Theta_X}$$

$$\frac{X_2 - \Theta_X}{T_2 + \Theta_X} \leq \frac{X_1}{T_1} \leq \frac{X_2 + \Theta_X}{T_2 - \Theta_X}$$

where $T_i = B_i + G_i + R_i$ $(i=1,2)$, X denotes one of B,
G, and R, and Θ_X (X=B, G, R), denote the threshold
values used for segmentation.

(2) The ratio between the widths of the candidate region
and the road is between 1/2 and 2.

(3) The distance between the road and one end point of
the longest path of the candidate region is less than
$3W_1$, where W_1 denotes the width of the candidate re-

region.

Thus, once a road is recognized, the RD2 subsystem tries to
bind a side road with the recognized road. By iterating this operation, RD2 can recognize a complex road network. Figure 6.7 shows
the roads recognized by the RD2 subsystem, where side roads are
connected to a main road and recognized as one road. While the
disjoint regions on the upper left side are recognized as one road
in the blackboard, they are separate in the two-dimensional space
because the elementary regions between them cannot be included
into the road.

The subsystems for the recognition of rivers, RV1 and RV2,
perform almost the same analysis processes as the subsystems RV1
and RV2, respectively. The differences from the road detection subsystems are:

(1) The candidate regions for rivers are water regions
(the RV2 subsystem does not require a candidate
region to be an elongated region).

(2) As described in Section 5.5, the multispectral properties of water are greatly affected by shadow.
Therefore, the RV2 subsystem merges shadow regions
with water regions which satisfy the following conditions:

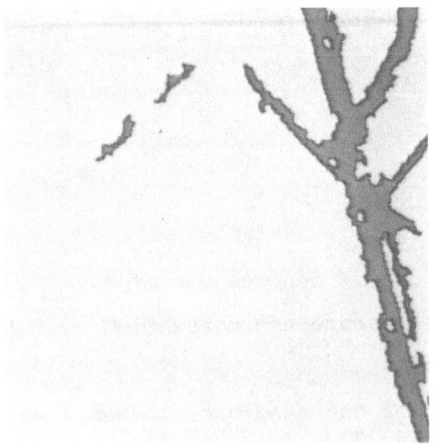

Fig. 6.7
Roads recognized by the RVI
subsystem; each road consists
of several elementary regions.
Two large side roads are connected to a main road and recognized as one road.

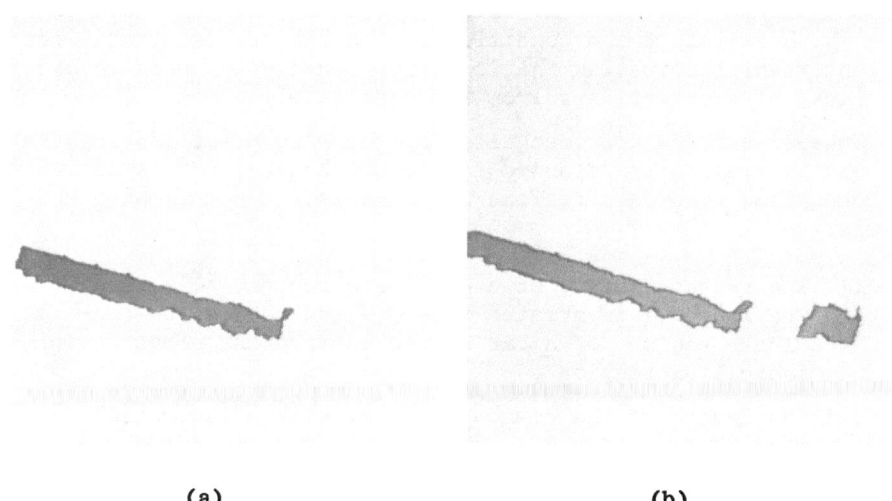

(a)   (b)

Fig. 6.8   (a) River recognised by the RVI subsystem in the picture
               of Fig. 8.3(c)

           (b) River recognized by the RV2 subsystem in the picture
               of Fig. 8.3(c); several shadow regions are merged
               with water regions

(a) adjacent to a water region:

(b) not vegetation region:

(c) if merged with a water region, the overall
    elongatedness increases.

Figure 6.8 (a) and (b) shows rivers recognized by RV1 and RV2,
respectively, in the picture of Fig. 8.3(c). Some water surfaces
in shadow which were not extracted as water regions are correctly
recognized as a river.

6.4.   Car

From the spatial and spectral characteristics of a car, we can
characterize it as having a rectangular shape and satisfying the
following logical expression.

$$(\text{large homogeneous region}) \wedge (\overline{\text{vegetation region}})$$
$$(\overline{\text{water region}}) \wedge (\overline{\text{shadow region}})$$

However, there exist a lot of false cars among the candidate regions which satisfy these conditions. The CAR subsystem utilizes
the contextual information that cars are usually on roads in order
to select only correct cars from the candidate regions.

Once some roads are recognized by the road-detection subsystems and the result is returned to the blackboard, CAR is activated
and recognizes candidate regions which satisfy the following conditions:

    (1) The rectangularity of a candidate region, FIT, (see
        Section 4.4) is greater than 0.7 (of course the can-
        didate region for a car should satisfy the above
        logical expression).

    (2) The length of the common boundary with a recognized
        road is more than 80% of the total region boundary.

Figure 6.9 shows the cars recognized by this method. We can
see that by using this environmental information, cars, which are
very difficult to locate, are successfully detected.

6.5. House and Building
6.5.1. HOUSE1

Generally, houses in an aerial photograph may be characterized by the following properties: rectangular shape, appropriate
area size, not vegetation region, and not water region. But with
these conditions alone, many regions are recognized as houses
which are not real houses as in the case of car recognition.
Houses do not have such distinguishing properties as elongatedness
for roads and rivers, and vegetation for crop fields, forest areas,
and grasslands. Thus, in order to locate real houses successfully,
the HOUSE1 subsystem first extracts rough areas where houses are
highly likely to exist, that is. residential areas, and then searches for houses only in the residential areas. This is the basic
idea of the focusing of attention. It will enable us to keep
"false houses" from being recognized as well as to save processing
time.

The HOUSE1 subsystem first tries to extract the candidate
areas for residential areas by the following method, assuming that
a residential area consists of a number of small objects such as
houses, gardens, roads, and shadows.

Step 1: Extract areas which satisfy

(high contrast texture area) ∧ ($\overline{\text{large homogeneous region}}$) ∧
($\overline{\text{large vegetation area}}$).

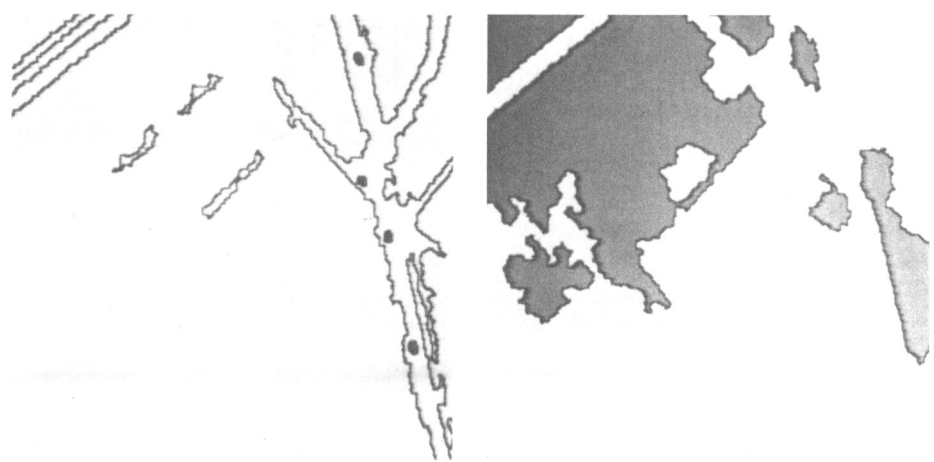

Fig. 6.9 Recognized cars   Fig. 6.10 Candidate areas for
                                              residential areas

Step 2: Select large areas among the areas extracted in
         Step 1, and make them the candidate areas for
         residential areas. (The threshold for area used
         here is the same as that used for large homoge-
         neous regions.)

Figure 6.10 shows the areas extracted by this method.

Then HOUSE1 selects the residential areas from these areas by
using the following knowledge: since in a residential area houses
are arranged systematically, the texture in it shows two prominent
directionalities which are at right angles to each other because
of the straight sides of the rectangular silhouettes of houses.
Thus HOUSE1 calculates the directionality of the texture in each
candidate area for a residential area by the following method.

Step 1: Differentiate the brightness picture by the ope-
         rators shown in Fig. 6.11, where each pixel is
         given the average brightness of the elementary
         region to which it belongs. (Here the bright-
         ness is the average of the gray levels in the
         four spectral bands.)

Step 2: Let $\Delta_x f(i, j)$ and $\Delta_y f(i, j)$ denote the differ-
         ential values at a point (i, j) calculated by
         these two operators, respectively. Calculate
         the direction of the gradient 0 at a point whose

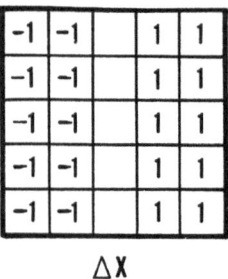

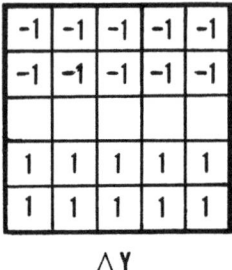

Fig. 6.11 Operators for differentiation

gradient magnitude is large by

$$\theta = \tan^{-1} (\Delta_y f(i, j) / \Delta_x f(i, j)),$$

where θ is rounded off into one of $0°$, $10°$,
..., $170°$.

Step 3: Make a histogram of the direction codes, $h_D(i)$
$(i = 0$ to $8)$, in the candidate area, where the
same direction code is given to those points
whose directions differ by $90°$, that is, 0 for
$0°$ and $90°$, 1 for $10°$ and $100°$, and so on.

Step 4: If

$$\max_{0 \le i \le 8} h_D(i) \ge 1.5 \left(\sum_{i=0}^{8} h_D(i) / 9 \right)$$

is satisfied, then recognize the candidate
region as a residential area.

Figure 6.12 shows the residential areas extracted by this
method.

The HOUSE1 subsystem searches for houses only in these residential areas. It uses very strict conditions to avoid misrecognition, since the properties of recognized houses will be used to
locate other houses which are outside of the residential areas or
deviate from the typical house. Conditions to be checked for the
recognition of houses are:

(a) in a residential area

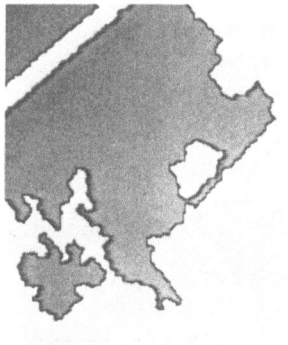

Fig. 6.12 Residential areas

(b) neither vegetation, shadow, nor water region

(c) shadow-making region

(d) the rectangularity, FIT (see Section 4.4), is great-
    er than 0.8.

HOUSE1 examines these conditions for each elementary region only
in residential areas. Fig. 6.13 shows the houses recognized by the
HOUSE1 subsystem. Even though all the houses could not be located
because of the strict conditions, no "false houses" have been recognized.

    Unrecognized houses will be located by other house detection
subsystems (HOUSE2, HOUSE3, and HOUSE4) which utilize the spectral
and spatial characteristics of recognized houses to search for new
houses. This feedback analysis, that is, to find very prominent
objects and then use their properties to recognize ambiguous objects, is quite useful for the recognition of objects without clear
distinguishing characteristics.

6.5.2. HOUSE2

    As one can see in Fig. 6.13, about half of the houses in the
residential areas are not recognized by the HOUSE1 subsystem.
This is because the conditions for houses used in HOUSE1 are made
very strict to avoid "false houses", that is,

(a) As the areas of houses are rather small, FIT is
    sometimes lowered by the noise on the boundary.

(b) Some of the houses do not make shadows on the ground
because the spaces between them are very narrow. As
a result, the regions corresponding to these houses
have not been regarded as shadow-making regions.
(See the blue houses in Fig. 2.1 aligned on the second
row in the large residential area.)

In order to keep the result of the analysis reliable, however, we
cannot relax these conditions.

We often encounter problems of this kind in picture processing. For example, when we try to extract edges in a picture, we
usually apply thresholding to the differentiated picture. If we
set the threshold very high, we miss some of the real edges, while
a lot of noise is extracted with a low threshold. It is very difficult to solve this problem by relying only on picture processing
techniques.

Shirai [69] showed that a heterarchical analysis is very
useful for detecting edges in pictures of polyhedra, where very
prominent edges are detected first and then the ambiguous ones are
located by using the properties of the already detected edges.
Since we can limit the candidates by using their relationships with
recognized edges, the threshold can be lowered without causing serious errors.

The idea of the heterarchical analysis (or the feedback analysis) has been incorporated in our system to recognize houses. The
HOUSE2 subsystem utilizes the spectral properties of already recognized houses to locate new houses. The HOUSE3 and HOUSE4 subsystems use the locational information of recognized houses.

Generally speaking, the spectral properties of objects tend to
change from picture to picture. However, it may be true to assume
that the objects of the same kind show quite similar spectral characteristics if they are in the same picture. Thus, the spectral
properties of recognized houses are quite useful for selecting the
candidate regions for houses, and hence we can relax the conditions
used in HOUSE1.

The recognition processes of the HOUSE2 subsystem are as
follows.

Step 1: Extract those elementary regions whose average
gray levels in the four spectral bands are similar to those of any already recognized house.
(The differences of the gray levels in the four
spectral bands are smaller than θ_i (i = B, G, R,
IR), respectively; θ_i denotes the gray-level
similarity measure used in segmentation.)

Fig. 6.13
Houses recognized by the HOUSE1
subsystem; since the conditions
are very strict, about half of
the houses are left unrecognized

Step 2:   Examine the following conditions for each can-
          didate region selected in Step 1, and if all
          conditions are satisfied, recognize it as a
          house:

          (a) neither vegetation, shadow, nor water region

          (b) not elongated region

          (c) FIT is greater than 0.7 (threshold is
              lowered

          (d) shadow-making region unless the region is
              included in any residential area.

Figure 6.14 shows the houses newly recognized by the HOUSE2
subsystem using the spectral properties of the houses in Fig. 6.13.
Many houses which have not been recognized by HOUSE1 have now been
recognized successfully.

6.5.3. HOUSE3

Ordinarily, a house has several (often two) roofs of different inclinations. These roofs are sometimes divided into different regions in segmentation because of shading and highlight.
The HOUSE1 subsystem tends to recognize the roofs in the shade because they are adjacent to shadow regions and thus are characterized as shadow-making regions. Since the spectral properties of

Fig. 6.14
Houses newly recognized by the
HOUSE2 subsystem using the
multispectral properties of the
already recognized houses

roofs in the sun and shade often differ very much (Fig. 6.15), the
HOUSE2 subsystem often cannot recognize the roofs on the sunny side.

The HOUSE3 subsystem tries to recognize unrecognized roofs
adjacent to already recognized roofs, and merge them into one house.
The recognition processes in HOUSE3 are as follows:

Step 1: Extract those regions which have long common
boundaries with already recognized houses.
(More than a quarter of the total boundary
points are adjacent to a house.)

Step 2: Check the following conditions for each candidate region, and if all conditions are
satisfied, then merge it with the adjacent
house and recognize the merged region as a
house:

(a) neither vegetation, shadow, nor water region

(b) FIT is greater than 0.7

(c) the ratio of areas between the candidate
region and the adjacent house is between
1/2 and 2.

(d) the hue of the region is similar to that of
the adjacent house. (The similarity of hue
is checked by the same method as in the recognition of roads.)

Since the gray levels in the four spectral bands of regions in the sun and shade are quite different, we utilize hue to check the similarity of the spectral properties. Figure 6.16 shows the houses recognized by the HOUSE3 subsystem, where two roofs of a house are merged into one region.

6.5.4. HOUSE4

In Japan, as shown in Fig. 2.1, houses in newly developed residential areas are arranged very systematically. The HOUSE4 subsystem locates unrecognized houses in residential areas by using the regularity in the arrangement of houses. The processes of the analysis in this subsystem are as follows:

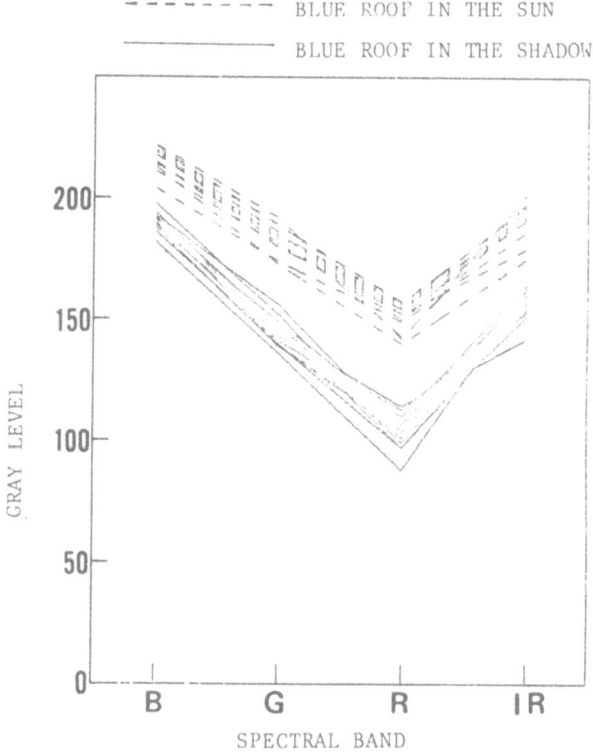

Fig. 6.15 Multispectral properties of roofs in sun and shade

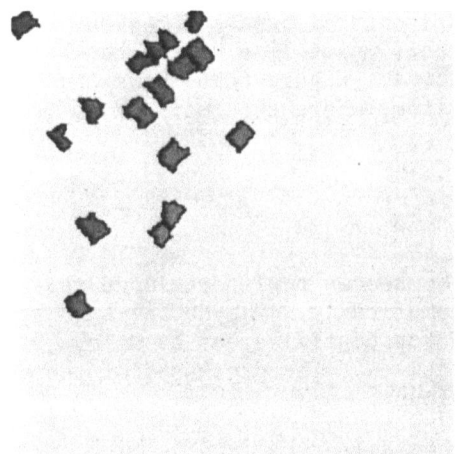 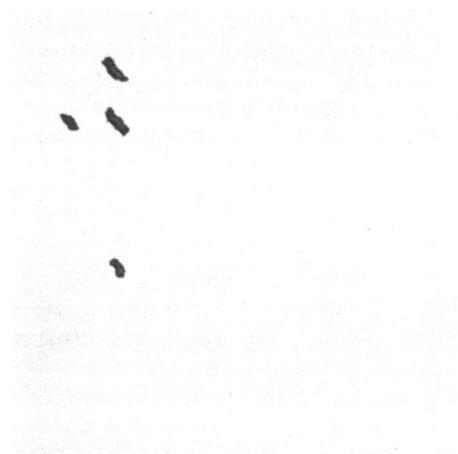

Fig. 6.16
Houses newly recognized by the
HOUSE3 subsystem; two adjacent
roofs of a house are merged into
one region

Fig. 6.17
Houses newly recognized by the
HOUSE4 subsystem; "missing houses"
in the large residential area are
correctly recognized

Fig. 6.18
Buildings located in the picture
of Fig. 8.3(c)

Step 1:   Extract already recognized houses in a residen-
          tial area. Then, find "regularity vectors" be-
          tween them by regarding each house as a texture
          element. (The detailed algorithm for extracting
          regularity vectors from relative vectors between
          texture elements was described in Section 3.3.3.)
          If no regularity vectors are extracted, then
          check other residential areas.

Step 2:   Let $\vec{v}_i$ (i = 1, 2, ..., n) denote the regularity
          vectors and $\vec{g}_j$ (j = 1, , ..., N) positions of the
          centroids of the already recognized houses in the
          residential area. Using the LABEL PICTURE, find
          elementary regions to which pixels located at
          $\vec{g}_j \pm \vec{v}_i$ belong (for all combinations of i and j).
          (Since each pixel in the LABEL PICTURE is labeled
          with a unique region number, it is very easy to
          find the elementary region to which the pixel be-
          longs.) Then regard these elementary regions as
          candidate regions for houses.

Note: This process for finding missing texture elements
          ("missing houses") is slightly different from that
          described in Section 3.3.4. This is because we
          want to extract as many candidate regions as pos-
          sible.)

Step 3:   Check the following conditions for each candi-
          date region and if all of them are satisfied,
          recognize it as a house:

          (a) neither vegetation region, water region, nor
              shadow region

          (b) not elongated region

          (c) FIT is greater than 0.8.

Figure 6.17 shows the houses newly recognized by this subsystem. "Missing houses" in the residential area have been successfully located.

The four house-detection subsystems described above work cooperatively to locate houses in various environments. Thus our
system can correctly recognize houses even if they do not have
any outstanding features.

6.5.5. Buildings

Generally it is very difficult to distinguish between buildings and houses. There may be no differences between these two objects. However, as is obvious from the algorithm of the HOUSE1
subsystem, the house-detection subsystems cannot recognize large
buildings, because they tend to be excluded from residential areas
and their shapes are not necessarily rectangular. Therefore, we
added a subsystem which locates such large buildings. The process
of the recognition of buildings is as follows:

Step 1: Pick up candidate regions for buildings which
         satisfy

$\overline{\text{(vegetation region)}} \land \overline{\text{(water region)}} \land$ (shadow-making region),

         and whose area sizes are greater than 400.
         (This threshold is experimentally determined.)

Step 2: Check the straightness of a region boundary for
         each candidate region, and if it has a straight
         boundary, recognize it as a building. (The al-
         gorithm for checking the straightness of a region
         boundary is the same as that used in the crop
         field detection subsystem.)

No buildings were recognized in the picture shown in Fig. 2.1
since all shadow-making regions are very small. Figure 6.18 shows
the buildings which were located in the picture shown in Fig. 8.3(c).

7. CONTROL STRUCTURE OF THE SYSTEM

As discussed in Section 2.3, we have adopted a production system as the software architecture of the system for the structural analysis of aerial photographs. Consequently, the system becomes highly modular, and we can easily implement the diverse knowledge required to describe the situation on the ground surface. Moreover, the efficiency and reliability of object recognition are greatly increased by the heterarchical control structure of the system.

The original production system merely gives the basic ideas of knowledge representation and control structure of the system. We have to design the exact data structure and the detailed control mechanism to make the system suitable for the analysis of complex · aerial photographs.

Each object-detection subsystem, as described in the previous chapter, performs a specialized knowledge-based processing task to recognize specific objects independently of the others. The system incorporates several mechanisms to organize these mutually independent subsystems and to control the overall processing flow of the analysis. It always monitors the contents of the blackboard, and manages it to get consistent results from the analysis.

There are three main roles to be performed by the (control) system:

(1) Conflict Resolution: When some elementary region is recognized by multiple object-detection subsystems as several different kinds of objects, the system solves the conflict between them.

(2) Correction of Segmentation Errors: Since the initial segmentation is based only on the multispectral properties of each pixel, the system tries to modify the segmentation result when some indication of

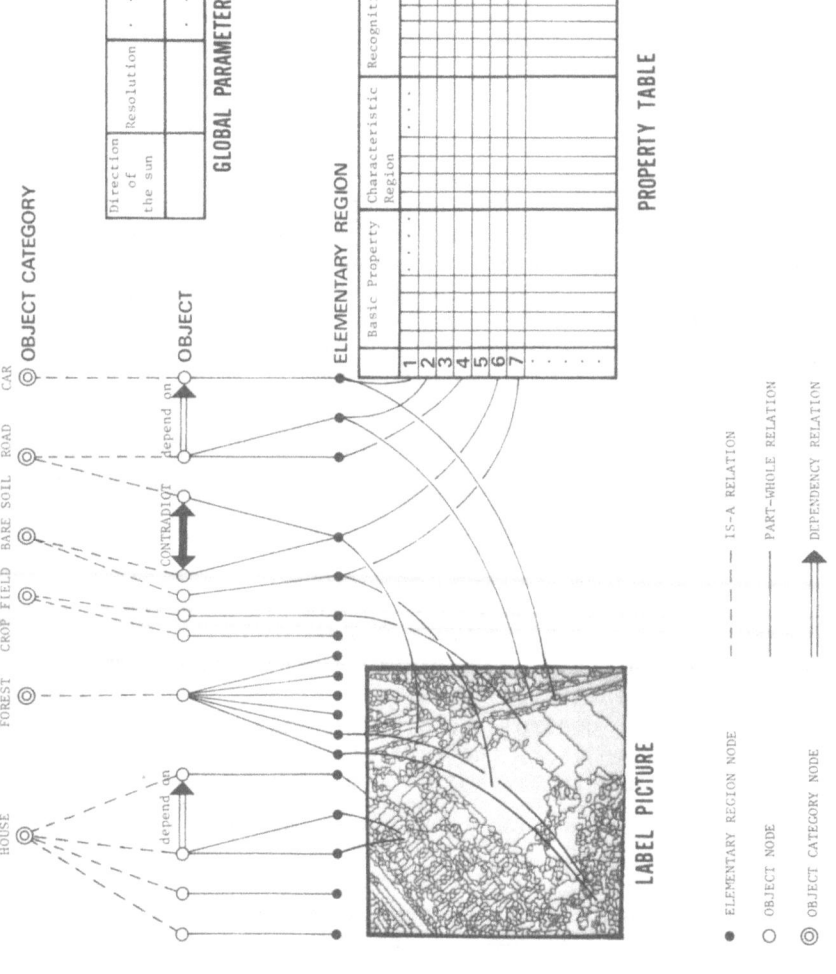

Fig. 7.1 Structure of the blackboard

segmentation error is suggested by an object-detection subsystem.

(3) The system checks the conditions to stop the analysis and to output the result.

With these general control mechanisms in the system, each object-detection subsystem can devote itself to its own processing without considering the results from the others, and it need not worry about minor errors in the initial segmentation.

In the following sections, we shall describe in detail the data structure of the blackboard and the control mechanism of the system.

7.1. Structure of the Blackboard

The blackboard is the sole place in the system for storing all the information to be recorded. It stores the observed facts and assertions about the scene under analysis, such as the properties of and the relations among regions and objects. Each object-detection subsystem interfaces with it in a uniform way to test the conditions for activation and to write in the result of the analysis. All the communications among subsystems are made via this blackboard, and there exist no private communication channels between object-detection subsystems. Thus, we can see the state of the processing by examining only the blackboard.

Resolution of a picture	Direction of the sun	whole picture	Average gray levels over the				properties for spectral Threshold values				Threshold value for area size
			B	G	R	IR	B	G	R	IR		

Fig. 7.2 The global parameter table

The data structure of the blackboard depends heavily on the nature of the application, which ranges from a simple collection of symbols to a complex graph structure. Figure 7.1 shows a schematic drawing of the structure of the blackboard in our system, which contains some tables, a symbolic picture, and a network. The network consists of three types of nodes, *i.e.*, "elementary-region nodes", "object nodes", and "object-category nodes". These nodes are organized in a hierarchical tree, and are mutually connected by several kinds of pointers, such as "part-whole relations", "is-a relations", and "dependency relations".

When an object-detection subsystem recognizes an object, it generates a new object node in the blackboard. The object node is connected with its constituent elementary regions by the pointers which denote the part-whole relations. (An object may correspond to one or more elementary regions.) Then the object node is linked with an object category node by an is-a relation. This pointer denotes that the object belongs to that category of object. For example, when houses A, B, and C are recognized, each of the object nodes corresponding to these houses is connected with the object category node for house.

If the recognition of an object is made by using the properties of the already recognized objects, the node representing the newly recognized object is linked with those of the old objects by the dependency relations. These pointers denote that the recognition of the former depends on that of the latter. For example, the recognition of a car always depends on that of roads since a car is recognized on the basis of its spatial relationships with the roads. When the recognition of some object is revoked by the conflict-resolution mechanism of the system, the objects which were recognized on the basis of the properties of the cancelled object are also cancelled by traversing dependency links. For example, if the recognition of a road is revoked, that of a car which was recognized using the properties of the cancelled road is also revoked. All nodes referred to cancelled objects are deleted from the blackboard and the analysis is re-started.

The blackboard contains three data organizations for storing a variety of information about regions and objects: the global parameter table, the property table, and the LABEL PICTURE (Fig. 7.1).

7.1.1. The Global Parameter Table

The global parameter table contains several numerical values which denote the overall properties of the picture data under analysis. Some of them denote the photographic conditions of the aerial photograph to be analyzed, such as the direction of the sun and the area size of a pixel on the ground. These parameters are given as

		RECOGNITION STATUS FIELD															CHARACTERISTIC REGION FIELD								BASIC PROPERTY FIELD										
REGION NUMBER	AREA SIZE	HOUSE4	HOUSE3	HOUSE2	HOUSE1	BUILDING	CAR	RV2	RRV1	RRD12	RD1	GRASS	FOREST	BS2	BS1	CF2	CF1	High Contrast Texture Area	Water Region	Large Vegetation Area	Vegetation Region	Shadow-Making Region	Shadow Region	Elongated Region	Large Homogeneous Region	BASIC SHAPE FEATURE DIREC	ELONG	FIT	AVERAGE GRAY LEVEL IR	GR	B	LOCATION BE Y XY	BB X YX		

Fig. 7.3 The property table

input data before the analysis. Others show the qualities of the
picture data, which are calculated by various picture processing
routines during the processes of segmentation and extraction of
characteristic regions. They include the average gray level in
each spectral band over the whole picture, the threshold values
used for checking the similarity of multispectral properties, and
the threshold of area size for specifying large regions (Fig. 7.2).

From these global parameters each object-detection subsystem
obtains information about the quality of the picture under analysis, and adjusts the parameters it uses in the recognition process.
Thus, it can successfully find objects in spite of unstable photographic conditions.

7.1.2. The Property Table

In our system the elementary regions, which are segmented
according to the multispectral properties by the segmentation process, are considered as the basic units for all higher-level processes. The property table stores various properties and recognition status of these elementary regions. Each row of the table
is allotted to one elementary region, and consists of three different fields: the basic-property field, the characteristic-region
field, and the recognition-status field (Fig. 7.3).

The first field is used to store various basic properties of
each elementary region which are calculated after segmentation
(see Section 4.4). They are quite general properties which characterize the multispectral properties, the shape, and the location of
each elementary region. These properties are used to extract characteristic regions and to recognize objects.

The results of the extraction of characteristic regions are
stored in the characteristic-region field of the property table.
Each characteristic-region extractor has a specific column in this
field where it (exclusively) returns the result of the analysis.
Most types of characteristic regions (large homogeneous, elongated,
vegetation, shadow, shadow-making, and water regions) can be denoted by simple "yes/no" flags in the corresponding columns, because each of them corresponds to a single elementary region. For
example, when an elementary region is extracted as a large homogeneous region, the column corresponding to the large homogeneous
region will be flagged. However, a characteristic region, such as
a large vegetation area and a high-contrast texture area, consists
of a group of elementary regions (see Fig. 5.12 and Fig. 5.15(d)).
In this case, each characteristic region of the same kind is first
numbered, and then the number of the characteristic region is
written in the corresponding column of its constituent elementary
regions (Fig. 7.4).

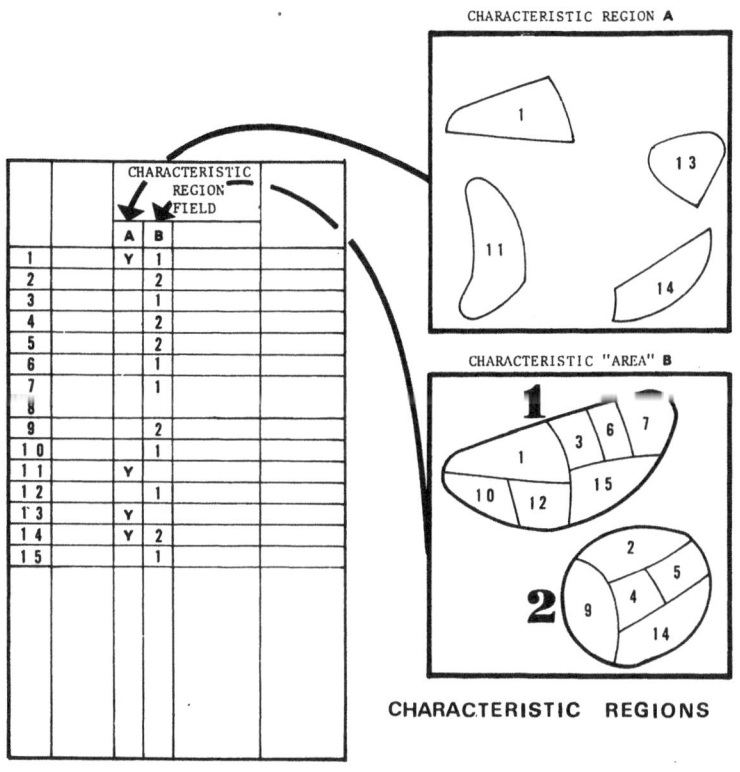

PROPERTY TABLE

Fig. 7.4 Representation of characteristic regions; characteristic
regions are represented in terms of various flags in the
characteristic-region field of the property table

Each characteristic-region extractor extracts specific characteristic regions independently of the others. Therefore, an elementary region is sometimes characterized as different kinds of
characteristic regions. For example, an elementary region can be
simultaneously a large homogeneous region and an elongated region
(see Figs. 5.2 and 5.3).

All characteristic regions are represented in terms of various
flags in the characteristic-region field of the property table, so
that we need not store their two-dimensional images in the blackboard. However, to extract spatial domains of objects such as residential areas, we need the two-dimensional images of the characteristic regions (see Section 6.5). We can easily generate them
by using the LABEL PICTURE: examine first the characteristic-region

field of the property table, and pick up constituent elementary regions of each characteristic region, then mark the pixels in the
LABEL PICTURE which belong to each extracted elementary region. In
this marking process we have only to scan within the rectangle area
enclosing an elementary region which is specified in the basic property field of the property table (see Section 4.4). Therefore it
does not take a long time to generate the two-dimensional images
of characteristic regions.

The recognition-status field in the property table consists of
many columns, each of which is exclusively used to store the recognition status returned from each object-detection subsystem (Fig.
7.3). (Thus, the number of columns in this field is the same as
that of object-detection subsystems.) The kinds of recognition
status flags in a column are:

(1) Unanalyzed: Neither recognized nor rejected. Ini-
    tially all the columns in the recognition status
    field for all elementary regions are set to this
    state.

(2) Rejected: Not an object to be detected. The possi-
    bility of the object is completely denied. An
    object-detection subsystem never analyzes an ele-
    mentary region having this flag in the correspond-
    ing column in the recognition status field. (Of
    course, the elementary region can be recognized as
    an object by object-detection subsystem of other
    kinds.)

(3) Irregular-shaped: The shape of the region is not
    suitable for the object while all the other fea-
    tures are satisfactory.

(4) Recognized: Recognized as an object.

After each object-detection subsystem has analyzed an elementary region, it returns one of the above recognition status flags
into its private column in the recognition status field of the elementary region. Thus, the result of the analysis by an object-
detection subsystem is highly restricted. Accordingly, the system
can grasp the current situation of the analysis by monitoring only
the recognition-status field of the property table, even if it does
not know the exact analysis processes of the object-detection subsystems.

The system always checks the recognition-status field of each
elementary region, and if an elementary region is recognized as two
or more different objects (that is, the recognition status field
contains multiple flags of "recognized"), it tries to resolve the
conflict. If an elementary region is marked as "irregular-shaped",
the system activates a split/merge program to correct a possible

segmentation error. (These mechanisms will be described in detail in the next section.

7.1.3. LABEL PICTURE

We store the LABEL PICTURE to denote the spatial characteristics of elementary regions such as shapes and the spatial relationships among the elementary regions. It is a symbolic picture, where each pixel in an elementary region is labeled with the same unique region number (Fig. 4.4). The reasons why we utilize the LABEL PICTURE to represent spatial relations instead of using pointers indicating relations such as "above", "adjacent", "included", etc., are:

(1) As aerial photographs are taken far from the ground, the world of aerial photographs can be considered as a two-dimensional world. That is, the objects on the ground surface do not change their shapes nor their structures even if the camera position is changed. Moreover the objects are rarely occluded by other objects, except by shadows. Therefore, in the analysis of aerial photographs, two-dimensional spatial relationships may be sufficient to represent the structure on the ground surface. (In the analysis of indoor and outdoor scenes, it is essential to organize regions into objects while considering their three-dimensional arrangements.)

(2) Object-detection subsystems, as described in Chapter 6, perform various picture processing functions to recognize objects. Therefore, it is desirable and sometimes crucial to retain the two-dimensional images of the regions. For example, in order to discriminate forest areas from grasslands, the subsystem for forest extracts mixed areas of shadow and shadow-making regions by applying shrinking and enlarging operations (see Section 6.2). These operations are truly two-dimensional picture processing functions which can hardly be performed except on two-dimensional images.

(3) The spatial relationships among regions used by object-detection subsystems are very diverse and depend on the properties of objects they want to find. Therefore, it is not economical, and is even sometimes impossible, to calculate all spatial relationships in advance. For example, one of the house detection subsystems (HOUSE4) estimates the locations of missing houses by using the regularity

in the arrangement of already recognized houses
(see Section 6.5). In this case, to pick up candidate regions for houses, processing on the two-
dimensional space is inevitable, and the regularity
of the arrangement cannot be represented by simple
pointers.

Since each pixel in the LABEL PICTURE contains a unique region
number, we can easily access the property table. On the other hand,
as mentioned in Section 4.4, the location of a region in the LABEL
PICTURE is denoted in the basic-property field of the property
table. We can also get the two-dimensional image of the region
quite easily.

7.2. Conflict Resolution

Each object-detection subsystem locates objects independently
of the other subsystems, so that an elementary region is sometimes
recognized by multiple subsystems at the same time. When multiple
recognition of an elementary region occurs, its recognition status
field contains multiple flags showing "recognized". The system
always checks the recognition status field of each elementary
region, and goes into the process of conflict resolution when it
finds evidence of multiple recognition.

This gives rise to two types of actions, depending on whether
or not the recognized objects to which the elementary region belongs are of the same category.

CASE 1: Multi-Category

When an elementary region is recognized as multiple objects
of different categories, the system evaluates the reliability value
of each object to which that elementary region belongs. The reliability of an object is calculated on the basis of the similarity
between the properties of the regions and those of the model of the
object. Which properties are incorporated in the calculation depends on the category of the object. For example, a region recognized as a road is given a larger value as its elongatedness increases. The system accepts the most reliable object and cancels
the recognition of less reliable objects; it removes the object
nodes except for the most reliable one and changes the corresponding recognition status of the elementary region from "recognized"
to "rejected". Since the object-detection subsystems do not analyze the regions marked as "rejected", the rejected objects are
never recognized again. This enables the system to avoid the infinite loop of recognition-and-conflict-resolution.

As an example of conflict resolution in the multi-category
case, suppose that elementary regions E_1, E_2, and E_3 as a whole

are recognized as the object O_1 of the category A by the object-detection subsystem SUB1, and E_3, E_4, and E_5 as the object O_2 of the category B by the object-detection subsystem SUB2 (Fig. 7.5). The system finds a conflict between SUB1 and SUB2 when examining the recognition status field of the region E_3, where two "recognized" flags are set by SUB1 and SUB2. Then it evaluates the reliability values of the objects O_1 and O_2.

Now, suppose that the reliability of the object O_1 surpasses that of O_2, then the object node of O_2 is removed from the blackboard and the recognition status of the region E_3 for the subsystem SUB2 is changed to "rejected". On the other hand, those of regions E_4 and E_5 are changed to "unanalyzed". The reason for this is as follows: the regions E_4 and E_5 do not exactly overlap with the object O_1 in the two-dimensional space. Therefore, the system reserves the possibility of these regions (E_4, E_5) being recognized by SUB2 again, while the region E_3 must not be analyzed by SUB2 because it will cause the same conflict. If E_4 or E_5 overlap with the object O_1, the recognition status is changed to "rejected".

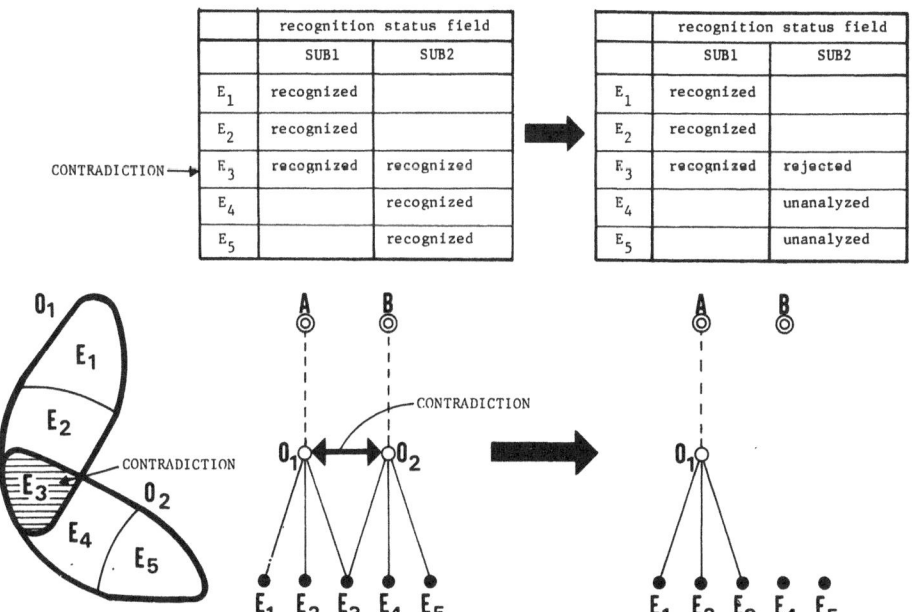

Fig. 7.5 Conflict resolution in the multi-category case (see text)

When the system deletes an object node as the result of conflict resolution, it checks the dependency links connected to that
object node. If there are objects which were recognized on the
basis of the properties of the object to be deleted, the system
also removes object nodes corresponding to such subordinate objects
by traversing the dependency links. For example, suppose that an
object O_2 was recognized based on the properties of an object O_1
and that the recognition of the object O_1 is cancelled (Fig. 7.6).
Then the system traverses the dependency link from the object O_1
and cancels the recognition of the object O_2. In this case, the
recognition status of the constituent elementary regions of the
object O_2 (E_1 and E_2 in Fig. 7.6) is given back to "unanalyzed".
Thus these elementary regions may be recognized again. This is because the recognition of the object O_2 itself does not cause any
contradiction, so that the system reserves the possibility of the
elementary regions E_1 and E_2 being recognized as an object. This
backtracking process enables the system to give back the contents
of the blackboard to the correct state.

In the course of the analysis of the picture shown in Fig.
2.1, a grassland in Fig. 7.7(a) causes conflicts with crop fields
(Fig. 6.2) and a forest area (Fig. 6.3) respectively. First the
conflict with a crop field was resolved and the overlapping region
was decided to be a crop field. Then, the rest of the area is recognized as a grassland again because the recognition status of
non-overlapping regions is given back to "unanalyzed" (Fig. 7.7(b)).
But this time, the newly recognized grassland comes to overlap with
a forest area and again it is rejected as the result of conflict
resolution. And the area shown in Fig. 7.7(c) is recognized as a
grassland once again. (This grassland also has a common area with
another crop field.)

Fig. 7.6 Removing a subordinate object by traversing a dependency
          link; when the object O_1 is removed from the blackboard,
          the object O_2 is also removed

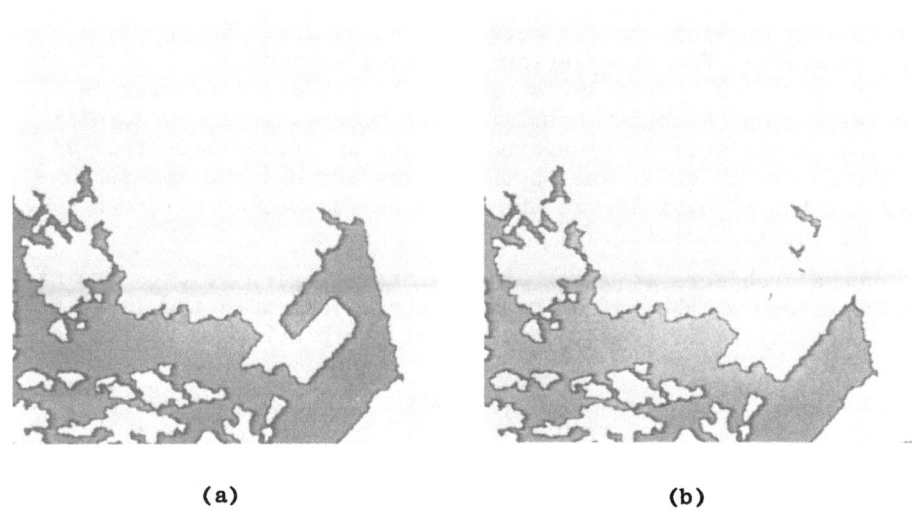

(a)   (b)

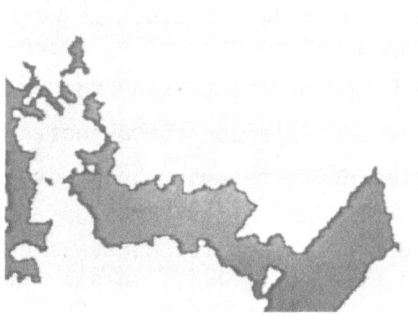

Fig. 7.7

(a) Grassland

(b) A newly recognized grassland
    after conflict resolution
    with a crop field in Fig. 6.2

(c) A newly recognized grassland
    after conflict resolution
    with a forest area in Fig. 6.3

(c)

CASE 2: Single Category

Since we have multiple object-detection systems for a single object category (for example, the present system contains two subsystems for the recognition of roads), an elementary region can be recognized by multiple object-detection subsystems for the same object category. For example, an elongated region alone can be recognized as a road by the RD1 subsystem if its elongatedness is quite large. On the other hand, the RD2 subsystem, independently of RD1, connects several elongated regions and recognizes them as a single road. In this case, both columns for RD1 and RD2 subsystems in the recognition status field may be marked as "recognized" at the same time.

When all objects to which an elementary region belongs are of the same category, the system prefers the object with the largest area. Then, it modifies the contents of the blackboard in the following way:

Suppose that an elementary region E_1 is recognized as an object O_2 of a category A, and that elementary regions E_1 and E_2 together are recognized as an object O_1 of the same category as O_2 (Fig. 7.8). In this case, since O_1 is larger than O_2, the system justifies O_1. Then the system removes the "is-a link" between the object category A and the object O_2. The reason why we do not delete the object node for the smaller object (O_2) is as follows: It may happen that the larger object O_1 comes to share some elementary region, say F_0, with another object of a different category (O_3) and be deleted from the blackboard as a result of conflict resolution (Fig. 7.9). Then the "suspended" object, that is, O_2 comes to be connected to the object category node A again. If we had removed the object node O_2 and changed the corresponding recognition status of the elementary region E_1 from "recognized" to "rejected", we could not recognize the object O_2 again; the object-detection subsystems never recognize elementary regions whose corresponding recognition status has been marked as "rejected".

This type of conflict resolution often happens for houses and roads as shown in Figs. 6.13, 6.16, and Figs. 6.5, 6.7. In all these cases, the largest objects are retained in the final result of the analysis. That is, smaller houses and roads are certainly present in the blackboard, but they are not authorized by the system, that is, they cannot be accessed from the object-category nodes. Thus we can count the number of houses and roads correctly by enumerating the number of object nodes connected to the object-category nodes.

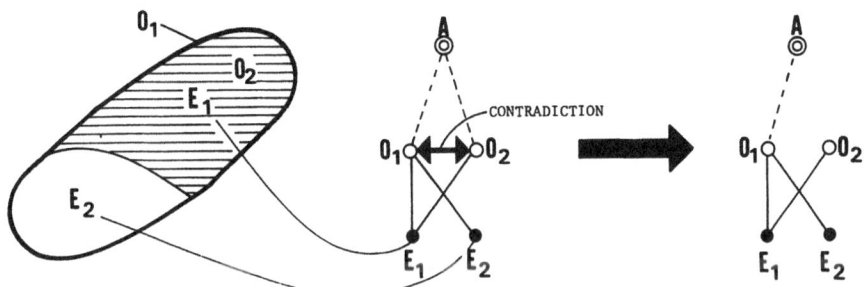

Fig. 7.8   Conflict resolution in the single-category case (see
           text)

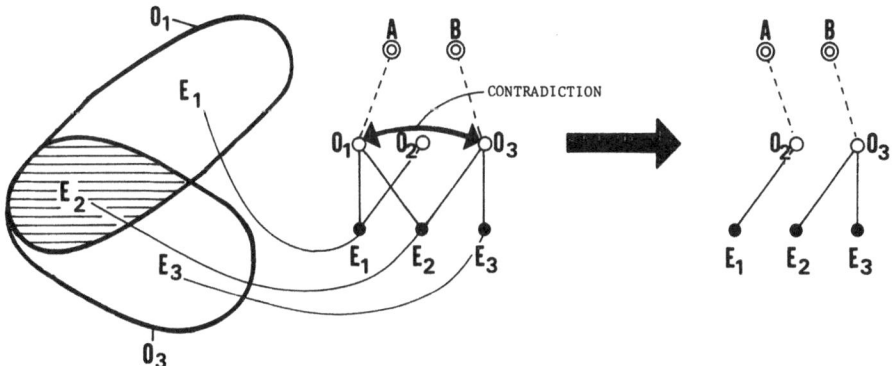

Fig. 7.9   Re-authorization of a "suspended" object (see text)

7.3. Correction of Segmentation Errors

The initial segmentation is made by relying only on the multispectral properties of pixels, which may sometimes cause errors
such as merging of regions which actually correspond to different
objects into one elementary region, and splitting of a "real"
region representing an object into several small elementary regions.
Thus the elementary regions do not always correspond to real objects. These errors in initial segmentation are repaired by the
system in the course of the recognition process, on the basis of
suggestions made by object-detection subsystems.

Each object-detection subsystem checks various properties of
elementary regions to recognize objects. It returns the recognition-
status flag, "irregular-shaped", when the shape of an elementary
region under analysis is not suitable for the object to be detected
while all the other properties are satisfactory. When the system
finds an elementary region marked as "irregular-shaped", it tries
to modify the segmentation and to generate one or more new regions
by applying a split/merge program.

First, the system assumes that the irregular-shaped elementary
region is the result of mismerging different regions and that the
mismerging causes one or more "bottle-necks" at the joint(s) of
these regions. (See Fig. 7.10, where regions R1 and R2 are merged
into one elementary region.) And the system tries to split the
elementary region into some smaller regions. The assumption might
not always be true. However all that we can do to correct the mismerging is to examine the shape of the region. We have no other

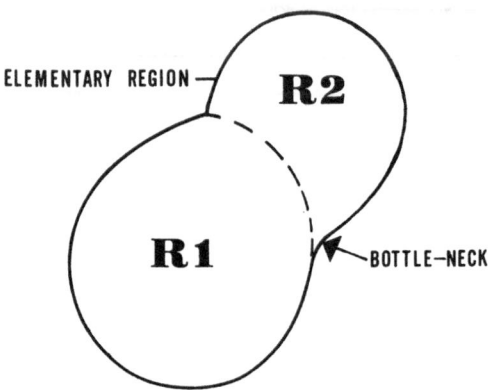

Fig. 7.10   An irregular-shaped region with a "bottle-neck"; two
            regions R1 and R2 are mismerged into one elementary
            region. Here we use "bottle-neck" to denote a con-
            stricted part of a region

way of finding proper edges with the region where it could be split. That is because our analysis is not a strong top-down process; the system cannot segment the elementary region on the basis of information about shapes of supposed objects which is transferred from the top-level analysis process. Therefore, if different regions have been merged into one without causing any "bottle-necks", the system cannot correct the segmentation error.

The algorithm to split an irregular-shaped region at "bottle-necks" is as follows:

Step 1: Calculate the change of width of the region along the longest path on the skeleton (Fig. 7.11). (For the longest path on the skeleton, see Section 3.2).

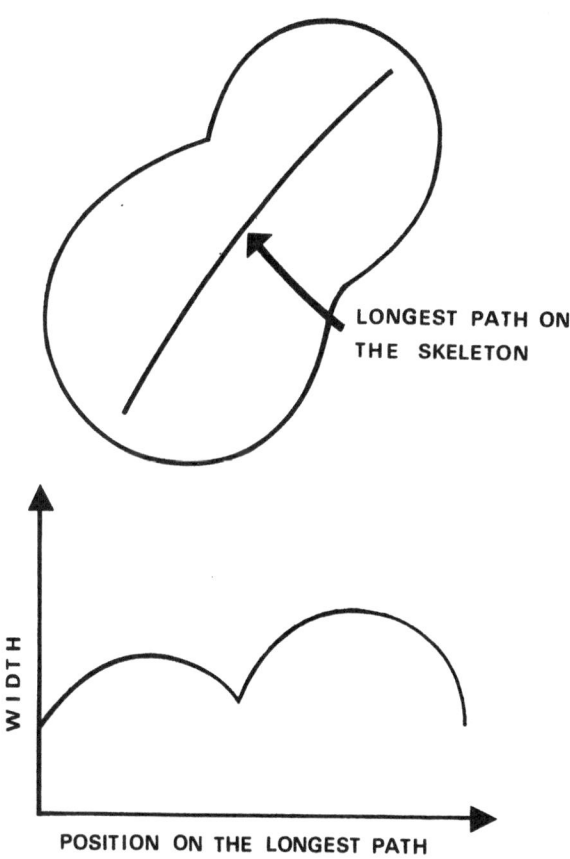

Fig. 7.11 Finding a "bottle-neck" of an irregular-shaped region

Step 2: Find valleys in the graph of the change of width
         and divide the region into several small regions
         at the positions corresponding to the valleys.
         (For a detailed algorithm for valley detection,
         see box, p. 165.)

If the region has a "bottle-neck", the width across the bottle-neck
becomes quite narrow. This can be detected by finding a valley in
the width curve of the region along its longest path. Thus, by
using the above algorithm, the system can find "bottle-necks" of
the region.

When the irregular-shaped elementary region is split into
small regions, each new region is given a new as yet unused region
number and the LABEL PICTURE is modified to represent these new
small regions. Then, the original elementary region and the new
regions are connected by split links (Fig. 7.12). (These new regions are temporarily regarded as elementary regions.)

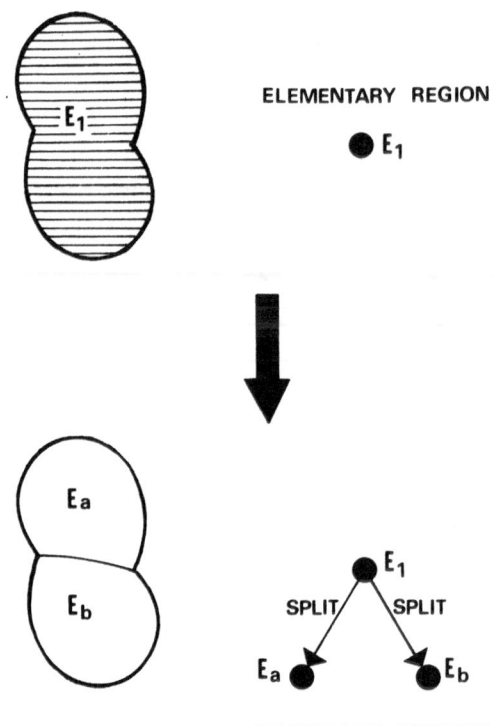

Fig. 7.12 Modification of the contents of the blackboard as the
           result of the split operation

"BOTTLE-NECK" DETECTION ALGORITHM

Let $W(i)$ $(i=1, 2, \ldots, L)$ denote a width of a region at the i-th position on the longest path on the skeleton.

Step 1:   Calculate the complement of $W(i)$,

$$\overline{W}(i) = \max_{i} W(i) - W(i) \quad i=1, 2, \ldots, L.$$

Set $I=1$ and $v_I=1$.

Step 2:   Starting from a position v_I, find a valley of the graph $\overline{W}(i)$, p_I, (that is, a peak of $W(i)$) by applying the valley-detection algorithm described in Section 4.3.2.

Step 3:   Starting from a position p_I, find a valley of the graph $W(i)$, v_I. $I=I+1$.

Step 4:   Repeat Steps 3 and 4 until the valley-detection algorithm comes across the end point of the longest path. Then, we have valleys (v_I) and peaks (p_I) $(I=1, 2, \ldots, N)$ in the graph $W(i)$.

Step 5:   Regard positions v_I as denoting those of bottle-necks of the region such that

$$1.5 \ W(v_I) \leq W(p_{I-1})$$

$$1.5 \ W(v_I) \leq W(p_I)$$

(We exclude the end point of the longest path, i.e., v_I.) If there are no valleys in $W(i)$ or no v_I which satisfy the above inequalities, the region is regarded as having no bottle-necks.

Since this algorithm uses the valley-detection algorithm for finding valleys and peaks in the graph of widths, it is insensitive to irregular noise on the region boundary.

If the irregular-shaped elementary region has no "bottle-necks" (no valleys in the graph), then the system activates the merging program. Neighboring small elementary regions with similar multi-spectral properties are merged with the irregular-shaped elementary region if the merging process increases the compactness of the merged region.

If neither splitting nor merging take place, the recognition status, "irregular-shaped", is changed to "rejected". When some neighboring regions are successfully merged with the irregular-shaped elementary region, a new region node which represents a merged region is generated, and each constituent region is connected with it by a merge link (Fig. 7.13). In this case no modification is performed on the LABEL PICTURE.

The new regions generated by the split/merge process are added to the blackboard as temporary regions; new elementary region nodes and new rows in the property table are generated for them, and the original regions and the new regions are connected by split/merge

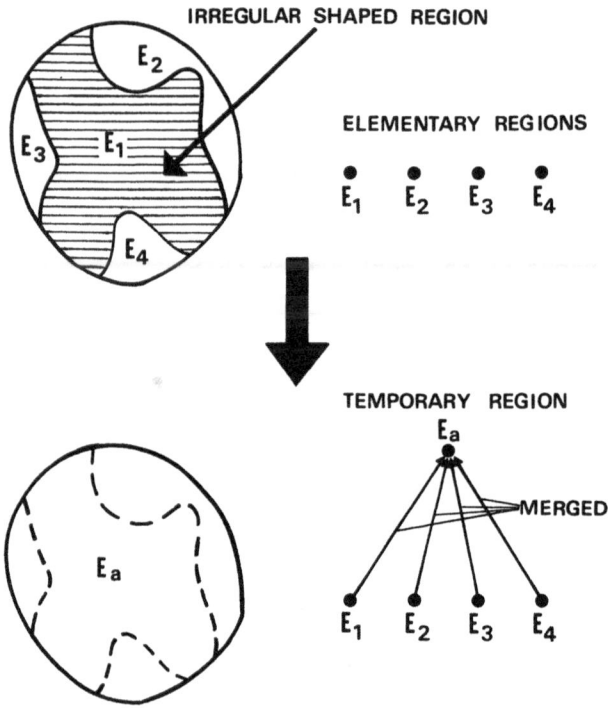

Fig. 7.13 Modification of the contents of the blackboard as the
result of the merge operation

links.   Then object-detection subsystems analyze them to attempt to
recognize objects.   If a temporary region is recognized as an object
by some object-detection subsystem and its result does not contra-
dict any recognition status of the original elementary region, then
the new region is registered as a real elementary region in connection with its related object, and the original elementary region is
deleted from the blackboard.   If the result of the recognition of
the new region contradicts that of the original one, the system
deletes one of them depending on the reliability.   When temporary
regions are not recognized as any objects at all, they are removed
from the blackboard, and the corresponding recognition status of
the original elementary region, which has been marked as "irregular-
shaped", is changed to "rejected".   Then, if the temporary regions
have been generated by splitting the original elementary region,
the LABEL PICTURE is repaired to represent the original elementary
region.

   In the case of the elementary region shown in Fig. 7.14(a),
the "bottle-neck" results from the error of mismerging two adjacent
houses.   Since the rectangularity of this elementary region is very
low, the house detection subsystem returns the recognition status

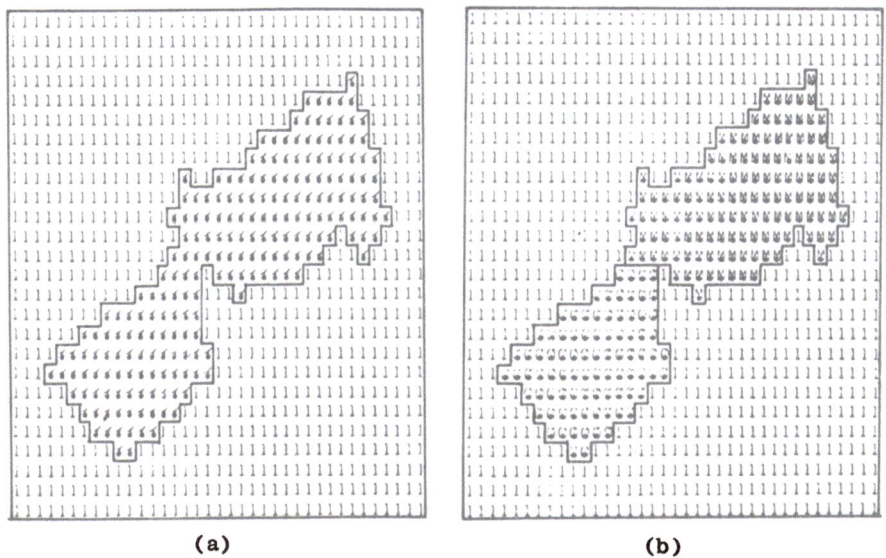

(a)   (b)

Fig. 7.14   (a) An irregular-shaped region with a "bottle-neck"; two
             real regions corresponding to houses are mismerged

            (b) Splitting the region shown in (a); each sub-region
             is recognized as a house

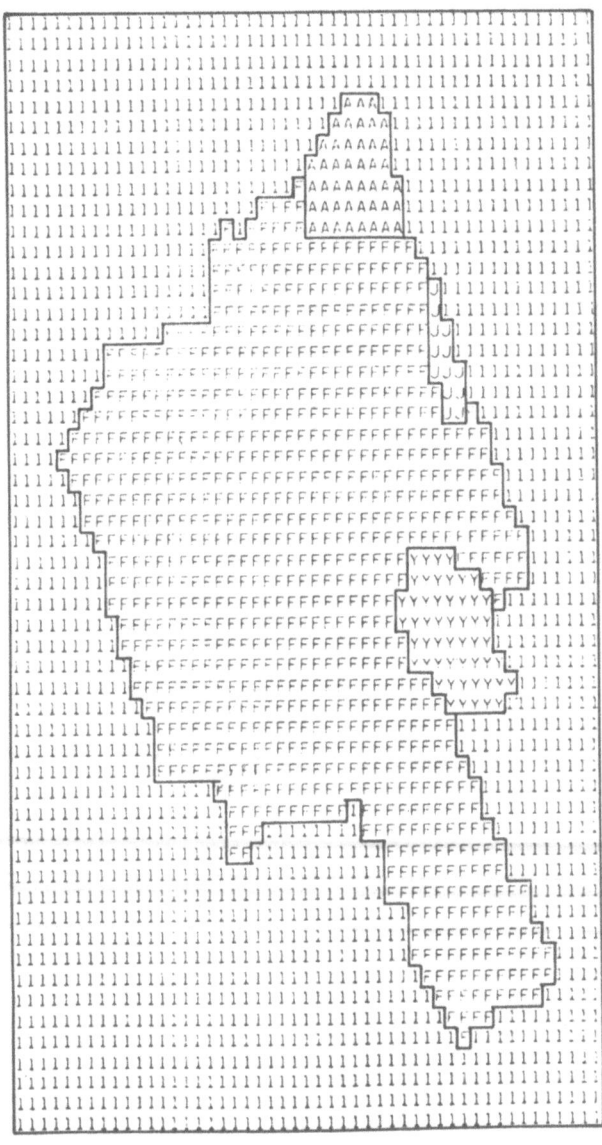

(a)

Fig. 7.15   (a) An irregular-shaped region (the central large region,
            in the picture shown in Fig. 8.3(a)).

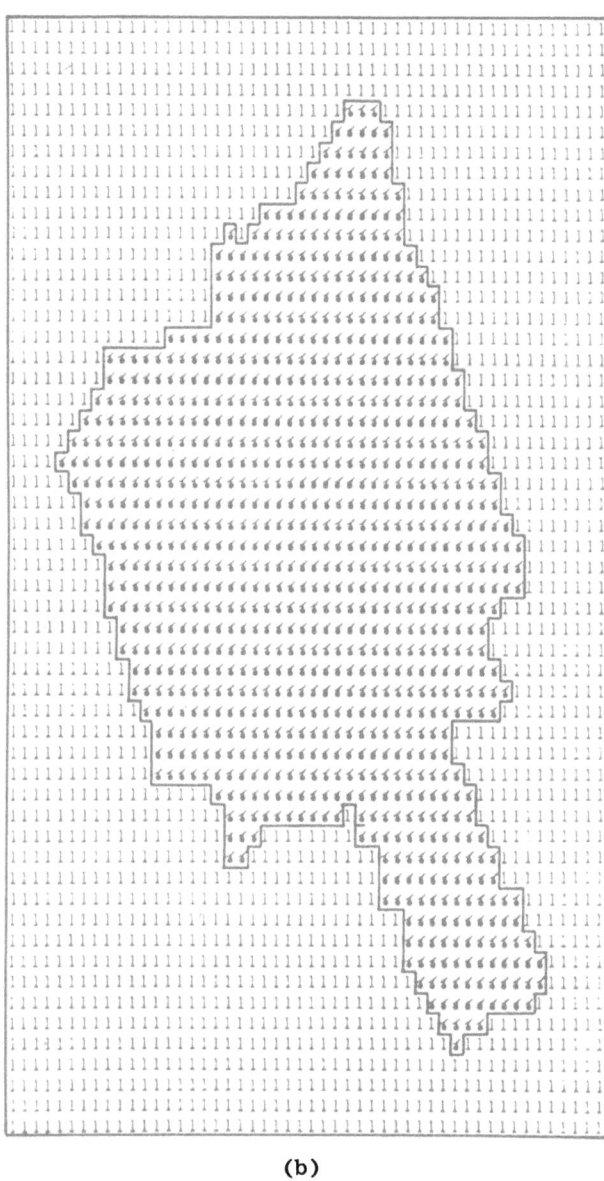

(b)

Fig. 7.15 (b) Result of merging neighboring small regions; the merged region is recognized as a crop field.

"irregular-shaped" to the recognition status field of this elementary region. When this elementary region is split into two separate regions (Fig. 7.14(b)), the rectangularity of each small region
becomes greater than the threshold, and hence the house detection
subsystem comes to recognize them successfully. As a result, the
node representing the original elementary region is removed from
the blackboard, and the corresponding row in the property table is
deleted.

Figure 7.15(a) shows an elementary region marked as "irregular-
shaped" by the crop-field detection subsystem because its boundary
is less straight. The system first tried to split this region.
However, it failed since the elementary region has no "bottle-necks".
Then the system tried to merge the elementary region with small adjacent elementary regions with similar multispectral properties to
create a region with a more compact boundary (Fig. 7.15(b)). The
straightness of the boundary of the merged region is increased, and
the region is successfully recognized as a crop field.

Due to the above-mentioned abilities of the system, i.e., the
conflict resolution and the correction of segmentation errors, each
object-detection subsystem can analyze regions and use the properties of already recognized objects without considering the results
of recognition by the other subsystems. All the interactions among
subsystems are exclusively managed by the system and the system
stops the analysis and outputs the result when no new objects are
recognized.

8. PERFORMANCE EVALUATION

8.1. Some Examples of the Analysis

Figure 8.1[1] shows the final result of analyzing the picture shown in Fig. 2.1 (the ID number of this aerial photograph is C6-7). Each recognized object is labeled with the color denoting its object category (Fig. 8.2), and is enclosed by a black contour line. The white areas denote unrecognized areas. The number and area size (in per cent) of the recognized objects belonging to each object category are shown in Table 8.1.

Since we do not have the detailed ground truth data of the district, the recognition result was evaluated by visual inspection of the photograph. Almost all large objects that have outstanding characteristics (roads, crop fields, grasslands, and forests) have been correctly recognized. Some wild areas are misrecognized as houses (they are marked with "R" in Fig. 8.1). These "false" houses were located by the HOUSE2 subsystem based on the multispectral properties of other recognized houses (Section 6.5.2). The multispectral properties of some "real" houses and these wild areas are almost identical, which led the HOUSE2 subsystem to the misrecognition. Even though the HOUSE2 sometimes causes errors like this, it recognizes many real houses which cannot be located by the other house detection subsystems. This is a trade-off between the efficiency and the reliability of the analysis which we often encounter in various application fields.

While the feedback analysis using the multispectral properties sometimes leads the system to misrecognition, the locational information is almost always helpful for locating small objects with less outstanding characteristics. Based on the knowledge that a house has roofs of different inclinations, the HOUSE3 subsystem locates yet unrecognized roofs adjacent to already recognized roofs. The

---

[1]All the figures in this chapter are in the color insert.

171

HOUSE4 subsystem estimates locations of unrecognized houses in a
residential area by using the spatial arrangement among recognized
ones. Due to these subsystems, many "ambiguous" houses have been
successfully recognized. Moreover, cars have been correctly located on the basis of their spatial relationships with roads. Since
these objects (houses and cars) are very small and have no prominent characteristics, it would be impossible to recognize them correctly without the locational information. In general, the feedback analysis by the model-driven object-detection subsystems has
proved to be very effective for the analysis of complex aerial photographs.

The red regions in Fig. 8.1 show unrecognized shadow regions.
The recognition of shadowed objects is one of the most difficult
problems in the analysis of aerial photographs. As mentioned in
Section 5.4, vegetation regions can be correctly extracted even if
they are in shadow. Therefore, shadow regions with vegetation may
be recognized as crop fields, grasslands, or forests. Indeed many
shadow regions in Fig. 2.1 have been successfully recognized as
forests and grasslands. In addition, the road and river detection
subsystems can recognize small shadow regions as parts of roads and
rivers. When shadowed areas are very large and do not contain vegetation, however, the present system cannot locate objects under
shadow. Some roads in the residential area in Fig. 2.1 have been
left unrecognized due to large shadows on them. Since the multispectral properties are greatly affected by shadow, we cannot rely
much on them to estimate objects under shadow. In order to recognize shadowed objects, the information about the global structure
of a scene will be required.

There remain unrecognized areas other than shadow regions
(white areas in Fig. 8.1). Their total area amounts to about 20%
of the whole picture. These areas correspond to wild areas and
small desert lands in residential areas. It is very difficult even
for human beings to identify them as particular objects. There are
many areas of this kind on the ground surface and it seems to be
reasonable to leave such areas as unrecognized.

The analysis result of the picture shown in Fig. 2.1 is quite
satisfactory, although there are some errors and some unrecognized
objects. In order to evaluate the general performance of the system,
we have applied it to several different aerial photographs of urban
and suburban districts. Figure 8.3 shows the original pictures of
the aerial photographs used in the experiments. Their size is 256
× 256 and one pixel corresponds to 50 × 50 cm on the ground.
Figure 8.4 shows the results of segmentation. We can see that the
segmentation process worked fairly well for different pictures.
Table 8.2 summarizes the parameters used in the analysis of these
aerial photographs. Some of them were given as input data and the
others were automatically determined by the system to make the processing adaptive to the picture data. Picture processing routines

Table 8.1 Summary of the result of analyzing the picture of Fig. 2.1

	Bare Soil	Crop Field	River	House	Road
Number of objects	3	3	0	53	5
Area size (%)	14.52	8.72	0.0	11.20	12.34

	Car	Grassland	Forest	Building	Shadow (unrecognized)	Unrecognized Area
Number of objects	4	59	2	0	–	–
Area size (%)	0.08	17.80	12.74	0.0	3.26	19.34

Table 8.2 Parameters used in the analysis

		C6-7	C4-8	C4-7	C3-6	C1-2
Average gray levels over the whole pictures	B	149	135	157	159	71
	G	105	90	100	115	106
	R	127	114	120	104	90
	IR	150	145	125	99	125
Resolution(size of a pixel on the ground)(cm)		50x50	50x50	50x50	50x50	50x50
Direction of the sun		-85°	60°	155°	150°	-60°
Threshold values for segmentation	B	14	9	14	9	12
	G	12	17	28	10	12
	R	11	14	11	10	9
	IR	11	13	13	12	16
Number of elementary regions		1044	629	1349	1156	816
Minimum area for large homogeneous regions		520	680	440	520	440
Number of large homogeneous regions		14	18	14	11	19
Threshold brightness for shadow regions		106	87	91	90	65

Table 8.3 Summary of the results of the analysis

		C4-8	C4-7	C3-6	C1-2
Bare Soil	No.	8	0	2	5
	Area(%)	24.50	0.0	2.25	13.38
Crop Field	No.	9	3	1	15
	Area(%)	20.67	5.04	0.94	36.80
River	No.	0	0	1	0
	Area(%)	0.0	0.0	4.58	0.0
House	No.	29	91	75	10
	Area(%)	5.72	16.53	13.38	2.86
Road	No.	3	3	5	2
	Area(%)	4.37	17.03	13.98	4.22
Car	No.	0	1	5	0
	Area(%)	0.0	0.02	0.09	0.0
Grassland	No.	54	106	60	70
	Area(%)	19.28	14.61	6.66	18.54
Forest	No.	3	6	3	6
	Area(%)	8.19	5.68	4.32	3.79
Building	No.	0	0	2	0
	Area(%)	0.0	0.0	1.58	0.0
Shadow (unrecognized)	Area(%)	1.60	10.25	22.07	3.97
Unrecognized area	Area(%)	15.67	30.84	30.15	16.44

in the system worked very well due to this ability of the adaptive
parameter determination. Figure 8.5 shows the final results of the
analysis of these pictures. (For the correspondence between the
object categories and the colors, see Fig. 8.2.) Table 8.3 summarizes the numbers and area sizes (in per cent) of the recognized
objects in these aerial photographs. Since the structures of the
scenes of C4-8 and C1-2 are rather simple, the results of object
recognition are quite satisfactory. On the other hand, in the
picture of C3-6, which is a very complex downtown scene, unrecognized areas amount to about 30%.

Generally speaking, large objects with prominent characteristics such as crop fields, bare-soil fields, roads, rivers, grasslands, and forests are correctly recognized even if a scene is very
complex. Car recognition is also very reliable due to the incor-
poration of locational constraint. However, it is very difficult
to recognize houses correctly in a complex scene. Considering less
prominent characteristics of houses, we have devised four house detection subsystems in order to realize the effective detection of
houses. We can say that these subsystems have worked very well and
located many ambiguous houses with, of course, some recognition
errors. Errors are caused partly because the resolution of the
pictures is somewhat low and partly because our model of houses is
rather simple; there exist several houses which are not rectangular
and have many non-rectangular roofs (Figs. 8.3 (b) and (c)).

As mentioned before, shadows often obstruct the object detection. Especially in aerial photographs of complex urban districts,
the effect of shadow is prominent and hence the recognition rate
is lowered. In C3-6, a part of the large vertical road has not
been located due to a large shadow on it. In C4-7, some roads are
split into many small regions by shadow (Fig. 8.4(b)) and hence
have been left unrecognized. Since the picture data alone cannot
give the information useful for estimating objects under shadow,
contextual and semantic information about the global structure of
a scene should be incorporated.

In C3-6 some buildings are left unrecognized and others are
misrecognized as bare-soil fields. This is because the process of
estimating shadow-making regions has failed due to the complicated
arrangement of three-dimensional objects. The information of
heights of points in a picture, which will be calculated from range
data and stereo photographs, will be needed for locating three-
dimensional objects in complex urban districts.

The process of correcting segmentation errors has prove to
be effective, especially for locating houses, since neighboring
houses are sometimes merged into one region. Many houses and some
crop fields have been located by this mechanism.

In summary, all the analysis processes from segmentation to
object recognition have worked fairly well for various types of

Table 8.4 Summary of processing times

Processing Time (sec.)	C6-7	C4-8	C4-7	C3-6	C1-2
	200	197	252	180	209

aerial photographs. The system locates a variety of objects in
various environments by using many features calculated from picture
data and the diverse knowledge of the objects. Especially, by using
the contextual information, the system has succeeded in recognizing
objects without prominent characteristics such as houses and cars.
Although there still remains some room for improvement, the results
of the analysis may be very satisfactory.

8.2. Processing Time Evaluation

Our system has been implemented on a large computer FACOM M-200[1]
and all programs are written in FORTRAN. The total number of FORTRAN
statements in the programs is about 12,000.

The edge preserving smoothing takes about three seconds to process a picture of 256 × 256. Since we iterate the smoothing operation 20 times for each picture in the four spectral bands, the
smoothing process takes rather a long time.

Table 8.4 shows the processing time required to analyze each
picture shown in Fig. 2.1 and Fig. 8.3, where we exclude the time
used for smoothing. Even though the analysis may seem to take a
long time, the processing time would be far longer if we had not
implemented the focusing mechanism. Table 8.5 shows an example of
the effectiveness of the focusing mechanism. TIME denotes the processing time for the crop field detection subsystem (CF1) to calculate the straightness of region boundaries. In the picture of C6-7
the CF1 subsystem focuses only on four candidate regions among 1,044
elementary regions and scans within their enclosing rectangles; it
can save very much processing time compared to a uniform overall

[1]FACOM M-200 is one of the largest computers in Japan and is
superior to the IBM 3033 processor. The FACOM M-200 computer
system available at the data processing center of Kyoto University has three CPU-s and 16 M byte LSI memory (access time
200 nsec). The average instruction time is about 90 nsec.

Table 8.5   Effectiveness of the focusing mechanism; in the picture shown in Fig. 2.1,
there are only four elementary regions that satisfy the conditions of the
crop field detection subsystem. Due to the focusing mechanism, the processing time is very much reduced

SCAN AREA	CANDIDATE REGION	TIME(sec.)	EFFICIENCY
Enclosing Rectangle	Four regions which satisfy conditions for crop fields	0.02	1
The Whole Picture Area	Four regions which satisfy conditions for crop fields	0.14	7
Enclosing Rectangle	All elementary regions(1,044)	0.94	47
The Whole Picture Area	All elementary regions(1,044)	24.02	1201

processing. One can see that if the focusing mechanism were not incorporated, the analysis process would take a prohibitive time.

Since image understanding systems for complex natural scenes must perform a variety of picture processing, they potentially require long processing time. In order to perform the analysis at high speed, some special-purpose hardware is needed for picture processing. Since a large number of aerial photographs is routinely produced nowadays, the development of such hardware is essential for putting image understanding systems into practical use.

9. CONCLUSION

9.1. Summary

We have described the system for the structural analysis of complex aerial photographs. This system is an image-understanding system which automatically locates a variety of objects using diverse knowledge of the world. Several experimental results have shown that it works fairly well for various types of scenes of urban and suburban areas.

In Chapter 1, we pointed out several shortcomings in the ordinary statistical classification and target detection, and stressed the necessity of introducing artificial intelligence techniques into the analysis of remote-sensing imagery. Our system is one of the first that attempts to automate the process of photointerpretation of complex aerial photographs.

Automatic photointerpretation by computer includes many different aspects of the problem: picture processing, feature extraction, knowledge representation, and control structure. Several sophisticated picture processing techniques have been newly developed for structuring the raw picture data. They have helped the system to transform a two-dimensional array of observed data into a well-organized data structure. Many experiments were performed for extracting features useful for the discrimination of objects. These features have facilitated the reliable recognition of diverse objects in aerial photographs. The primary focus of this research has been on how to organize individual picture processing techniques and diverse knowledge sources in a flexible way. Several new ideas have been introduced for realizing an efficient and reliable analysis: the focusing mechanism for estimating approximate areas of objects, the production system architecture for representing the diverse knowledge, the feedback (heterarchical) analysis for locating context-sensitive objects, and the feedback correction of segmentation errors.

The main achievements of this research are summarized as follows:

PICTURE PROCESSING TECHNIQUES

(1) A new smoothing method named "edge-preserving smoothing" has been developed for segmentation preprocessing. It not only removes noise in uniform regions but also sharpens blurred edges between regions. The iterative application of this smoothing transforms a noisy blurred picture into a distinct picture while keeping details of region boundaries. Experimental results showed that this smoothing is quite effective, not only for artificially generated patterns, but also for complex natural scenes such as aerial photographs. The excellent characteristics of edge-preserving smoothing facilitate the extraction of homogeneous regions in the segmentation process.

(2) A valley-detection algorithm has been devised for automatic threshold determination. This algorithm in substance locates a valley in a histogram by taking the gradient of the histogram at each point. In order to adaptively determine parameters for processing, our system applies this algorithm to various kinds of histograms: a histogram of differential values for finding the similarity measure used for region growing, and histograms of brightness values and area sizes for extracting shadow regions and large homogeneous regions, respectively. This algorithm is also used for finding constricted portions of an irregular-shaped region. The adaptive parameter determination enables the system to perform the analysis stably in spite of changeable photographic conditions of aerial photographs.

(3) In Section 3.3, we proposed a method for the structural description of textures consisting of regularly arranged elements. This method first extracts regularity vectors from a set of relative vectors between elements. The regularity vectors give the useful information for locating missing elements which picture processing programs fail to extract. The structures of regularly arranged patterns are described as overlays of two-dimensional lattices generated by the regularity vectors. The process of finding missing elements was used for estimating locations of unrecognized houses in a residential area. Thus, our system has succeeded in incorporating the information of spatial arrangements among objects into the recognition process.

FEATURE EXTRACTION

(4) Elongatedness is the most useful shape feature for locating elongated objects in aerial photographs. The method developed here for calculating the elongatedness of a region first extracts the longest path on the skeleton. Then it measures the width of a region at each point on the longest path. Therefore, it can correctly measure the elongatedness even for a curved region. In addition, the longest path on the skeleton is very useful for connecting several elongated regions into an elongated object, and a graph denoting

change of widths is used for finding constricted portions of an irregular-shaped region.

(5) Generally, the multispectral properties of objects are very sensitive to the photographic conditions of aerial photographs, and the analysis relying only on these characteristics cannot give stable results. However, it is known that vegetation and water areas show rather stable multispectral characteristics. We have found that ratios between gray levels in different spectral bands are very useful features for characterizing the multispectral properties of these areas. Using these features, the system can correctly extract vegetation and water regions despite changeable photographic conditions. These stable multispectral characteristics facilitated the discrimination of objects very much.

(6) Aerial photographs under analysis do not "explicitly" contain any three-dimensional information of objects on the ground. (That is, no information about heights of objects is given.) We have utilized shadow regions in order to estimate regions representing three-dimensional objects. Several experimental results have shown that extracted shadow-making regions show good correspondence to real three-dimensional objects. Even though we cannot measure the heights of objects, this three-dimensional information is very useful for discriminating three-dimensional objects from two-dimensional ones.

CONTROL STRUCTURE

(7) A focusing mechanism has been developed in order to realize an efficient and reliable analysis. The analysis process is divided into two steps: the global survey of the whole scene and the detailed examination of local areas. At the global survey stage, several kinds of characteristic regions are extracted. They represent primary features of objects and specify approximate areas where specific objects are highly probable. Experimental results have shown that these features are very useful for isolating a variety of objects in aerial photographs. At the detailed examination stage, a group of object-detection subsystems focus their attention on restricted local areas and locate specific objects by applying specialized programs to the local areas. This focusing mechanism not only saves processing time but also raises the reliability of object detection. This function is useful especially for large complex pictures such as aerial photographs.

(8) We have introduced a production system as the software architecture of the system. The diverse knowledge required to describe the structure on the ground surface is individually implemented in a set of object-detection subsystems. Each of them represents a specialized method for locating specific objects, and works independently of the others without any direct interactions. Therefore, we can easily modify any subsystem without worrying about

unexpected side-effects. The highly modular architecture of the
system enables us to augment its performance evolutionally via
trial-and-error experiments. This is an important factor in imple-
menting systems for such task domains that lack unified theories.

(9) A feedback analysis has been implemented for locating
context-sensitive objects. Some object-detection subsystems locate
objects by using the multispectral properties and the locational
information of already recognized objects. Due to these subsystems,
houses and cars, which are very difficult to recognize by the intrinsic properties of regions alone, have been recognized successfully. Thus this feedback analysis raises the efficiency of object
detection without decreasing the reliability.

(10) The (control) system integrates a set of mutually independent subsystems and controls the overall process of the analysis
by managing the contents of the blackboard. All object-detection
subsystems interface with the blackboard in a uniform way for checking conditions of activation and for writing the results of the
analysis. The system takes full responsibility for the maintenance
of the blackboard. When the system finds conflicts among object-
detection subsystems, it resolves them, and if necessary, it re-
stores the contents of the blackboard to the old state in order to
remove the effects of errors. This mechanism of conflict resolution and backtracking enables the system to coordinate a group of
mutually independent object-detection subsystems.

(11) The system has a large feedback loop from the high-level
processing stage to the low-level processing stage. The control
system activates a split/merge program according to the suggestions
given by object-detection subsystems. This program examines regions
generated by the initial segmentation and if possible, corrects segmentation errors. This mechanism is essential for the analysis of
complex pictures, because no simple picture processing programs can
give completely error-free results. Experimental results have shown
that this feedback loop in our system works very well and that several objects were successfully located due to this error-correction
mechanism.

9.2. Areas for Future Work

All functions mentioned above have enabled the system to perform an efficient and reliable analysis of complex aerial photographs. We believe that we have been able to demonstrate a model
of automatic photointerpretation. This study will be a step to-
wards the development of research on both remote sensing and image
understanding. We admit that there are many parts to be improved
in our system. Main topics which will contribute to the great ad-
vancement of our system are:

(1) Introduction of map information: Our system assumes that no *a priori* information about structures of scenes is given. As mentioned in Section 1.4, however, map information serves as an approximate model representing situations on the ground surface. This information will greatly facilitate the analysis of aerial photographs. On the other hand, the result of analyzing photographs will serve to update the "old" map data to match current situations. Thus, the organization of map data base and photointerpretation systems is crucial for the development of composite landuse pattern monitoring systems.

(2) Characterization of specialists' knowledge: We have shown that the multispectral properties of vegetation and water areas can be correctly characterized despite changeable photographic conditions of aerial photographs and that these characteristics are very useful for object discrimination. In order to discriminate plants in crop fields and trees in forest areas, we have to characterize the knowledge which human specialists use to interpret aerial photographs. For this purpose, intensive experiments on the multispectral and textural characteristics of various objects are required.

(3) Introduction of three-dimensional information: Although our system can estimate locations of three-dimensional objects by using shadow, we cannot know the exact heights of objects. Moreover, this function does not work well for complex scenes such as those of urban areas, where various three-dimensional objects are arranged in a complicated fashion. Therefore, the three-dimensional information given by range data and stereo photographs will serve very much for discriminating objects in such complicated situations.

(4) Parallel processing: Image understanding systems include many picture processing processes: preprocessing, segmentation, and calculation of properties of regions and edges. These processes generally take much time especially for large pictures such as aerial photographs. Even though our system attempts to reduce processing time by the focusing mechanism, it still requires long processing time for routinely analyzing a large amount of aerial photographs. It will be necessary to design special hardware architectures which will perform picture processing in parallel.

In conclusion, we hope that this work will contribute to the further development of remote-sensing and image-understanding research.

BIBLIOGRAPHY[1]

1. Aggarwal, R.K. and **Wittenburg, T.M.**, Syntactic Recognition of Tactical Targets, *Proc. of a Workshop on Image Understanding*, Pittsburg, Nov. 1978, pp. 48-58.

2. Akin, O. and **Reddy, R.**, Knowledge Acquisition for Image Understanding Research, *Computer Graphics and Image Processing*, Vol. 6, 1977, pp. 307-334.

3. Bajcsy, R. and **Tavakoli, M.**, A Computer Recognition of Bridges, Islands, Rivers and Lakes from Satellite Pictures, *Proc. of Machine Processing of Remotely Sensed Data*, 1973, pp. 2A-54 — 2A-68.

4. Bajcsy, R. and **Lieberman, L.I.**, Computer Description of Real Outdoor Scenes, *Proc. of 2nd IJCPR*, 1974, pp. 174-179.

5. Bajcsy, R. and **Tavakoli, M.**, Computer Recognition of Roads from Satellite Pictures, *Proc. of 2nd IJCPR*, 1974, pp. 190-194.

6. Barrow, H.G., **Bolles, R.C.**, **Garvey, T.D.**, **Kremers, J.H.**, **Lantz, K.**, **Tenenbaum, J.M.** and **Wolf, H.C.**, Interactive Aids for Cartography and Photo Interpretation: Progress Report, October 1977, *Proc. of a Workshop on Image Understanding*, Palo Alto, Oct. 1977, pp. 111-127.

7. Barrow, H.G. and **Fishler, M.A.**, An Expert System for Detecting and Interpreting Road Events Depicted in Aerial Imagery, *Proc. of a Workshop on Image Understanding*, Cambridge, May 1978, pp. 155-156.

8. Bauer, M.E., Technological Basis and Applications of Remote Sensing of the Earth's Resources, *IEEE Trans. on Geoscience Electronics*, Vol. GE-14, No. 1, 1976, pp. 3-9.

9. Bernstein, R. ed., Digital Image Processing for Remote Sensing, IEEE Press, New York, 1978.

10. Binford, T.O. and **Brooks, R.A.**, Geometric Reasoning in ACRONYM, *Proc. of a Workshop on Image Understanding*, Palo Alto, April 1979, pp. 48-54.

11. Bolles, R.C., **Quam, L.H.**, **Fischler, M.A.** and **Wolf, H.C.**, The SRI Road Expert: Image-to-Database Correspondence, *Proc. of a Workshop on Image Understanding*, Pittsburgh, Nov. 1978, pp. 163-174.

12. Brice, C.R. and **Fennema, C.L.**, Scene Analysis Using Regions, *Artificial Intelligence*, Vol. 1, 1970, pp. 205-226.

---

[1]IJCPR: International Joint Conference on Pattern Recognition
IJCAI: International Joint Conference on Artificial Intelligence

13. Brooks, R.A., Goal-Directed Edge Linking and Ribbon Finding, *Proc. of a Workshop on Image Understanding*, Palo Alto, April 1979, pp. 72-78.

14. Carlton, S.G. and Mitchell, O.R., Image Segmentation Using Textures and Grey Level, *Proc. of the IEEE Conf. on Image Processing and Pattern Recognition*, Troy, N.Y., pp. 387-391.

15. Carlucci, L., A Formal System for Texture Languages, *Pattern Recognition*, Vol. 4, 1972, pp. 53-72.

16. Cheng., G.C., Ledley, R.S., Pollock, D.K. and Rosenfeld, A., eds., *Pictorial Pattern Recognition*, Thompson Book Company, Washington, D.C., 1968.

17. Davis, R. and King, J., *An Overview of Production Systems*, A.I. Lab. Memo AIM-271, Stanford Univ., 1975.

18. Duda, R.O. and Hart, P.E., *Pattern Classification and Scene Analysis*, John Wiley & Sons, 1973.

19. Feigenbaum, E.A., Buchanan, B.G. and Lederberg, J., On Generality and Problem Solving: A Case Study Using DENDRAL Program, *Machine Intelligence 6*, Edinburgh Univ. Press, 1971.

20. Fejes-Toth, L., *Regular Figures*, Pergamon Press, New York, 1964.

21. Fu, K.S., On the Application of Pattern Recognition Techniques to Remote Sensing Problems, Technical Report, TR-EE71-13, Purdue Univ., 1971.

22. Fu, K.S., *Syntactic Methods in Pattern Recognition*, Academic Press, New York, 1974.

23. Fukada, Y., Spatial Clustering Procedure for Region Analysis, *Proc. 4th IJCPR*, 1978, pp. 329-331.

24. Garvey, T.D. and Tenenbaum, J.M., On the Automatic Generation of Programs for Locating Objects in Office Scenes, *Proc. of 2nd IJCPR*, 1974, pp. 162-168.

25. Gates, D.M., Keegen, H.J., Schleter, J.C. and Weidner, V.R., Spectral Properties of Plants, *Appl. Opt.*, Vol. 4, 1965, pp. 11-20.

26. Gonzalez, R.C. and Wintz, P., *Digital Image Processing*, Addison-Wesley Publishing Company, 1977.

27. Hanson, A.R. and Riseman, E.M., *Preprocessing Cones: A Computational Processor*, Tech. Report 74C-1, Dept. of Computer and Information Science, Univ. of Mass., 1974.

28. Haralick, R.M. and Kelly, G.L., Pattern Recognition with Measurement Space and Spatial Clustering for Multiple Images, *Proc. of IEEE*, Vol. 57, 1969, pp. 654-665.

29. Haralick, R.M., Shanmugam, K. and **Distein, I.**, Textural Features for Image Classification, *IEEE Trans. on System, Man and Cybernetics*, Vol. **SMC-3**, 1973, pp. 610-621.

30. Haralick, R.M. and **Distein, I.**, A Spatial Clustering Procedure for Multi-Image Data, *IEEE Trans. on Circuit and Systems*, Vol. **CAS-22**, No. 5, 1975, pp. 440-450.

31. **Haralick, R.M.**, Statistical and Structural Approaches to Texture, *Proc. of 4th IJCPR*, 1978, pp. 45-69.

32. **Henderson, R.G.**, Signature Extension Using the MASC Algorithm, *IEEE Trans. on Geoscience Electronics*, Vol. **GE-14**, No. 1, 1976, pp. 34-37.

33. **Holmes, W.S.**, Automatic Photointerpretation and Target Location, *Proc. of IEEE*, Vol. **54**, No. 12, 1966, pp. 1679-1685.

34. **Horn, B.K.P.**, and **Bachman, B.L.**, Using Synthetic Images to Register Real Images with Surface Models, *Proc. of a Workshop on Image Understanding*, Palo Alto, Oct. 1977, pp. 75-83.

35. **Horn, B.K.P.**, Hill-shading and the Reflectance Map, *Proc. of a Workshop on Image Understanding*, Palo Alto, April 1979, pp. 79-120.

36. **Huang, T.**, Per-Field Classification of Remotely Sensed Agricultural Data, *Proc. 1970 Allerton Conference on Circuit and System Theory*.

37. **Justusson, B.**, Noise Reduction by Median Filtering, *Proc. of 4th IJCPR*, 1978, pp. 502-504.

38. **Kanade, T.**, *Computer Recognition of Human Faces*, Birkhauser Verlag, Basel and Stuttgart, 1977.

39. **Kanade, T.**, Model Representation and Control Structures in Image Understanding, *Proc. of 5th IJCAI*, pp. 1074-1082.

40. **Kelly, M.**, Edge Detection in Pictures by Computer Using Planning, *Machine Intelligence* **6**, Edinburgh, 1971, pp. 397-409.

41. **Kettig, R.L.** and **Landgrebe, D.A.**, Classification of Multispectral Image Data by Extraction and Classification of Homogeneous Objects, *IEEE Trans. on Geoscience Electronics*, Vol. GE-14, No. 1, 1976, pp. 19-26.

42. **Lesser, V.L.** and **Erman, L.D.**, A Retrospective View of the HEARSAY-II Architecture, *Proc. of 5th IJCAI*, 1977, pp. 790-800.

43. **Li, R.Y.** and **Fu, K.S.**, Tree System Approach for LANDSAT data Interpretation, *Proc. IEEE Symp. on Machine Processing of Remotely Sensed Data*, 1976, pp. 2A-10 — 2A-16.

44. **Lillesand, T.M.** and **Kiefer, R.W**, *Remote Sensing and Image Interpretation*, John Wiley & Sons, New York, 1979.

45. **Lu, S.Y.** and **Fu, K.S.**, A Syntactic Approach to Texture Analysis, *Computer Graphics and Image Processing*, Vol. **7**, 1978, pp. 303-330.

46. **Marr, D.**, *Early Processing of Visual Information*, A.I. Memo No. 340, A.I. Lab., M.I.T., 1975.

47. **Milgram, D.L.**, Results in FLIR Target Detection and Classification, *Proc. of a Workshop on Image Understanding*, Cambridge, May 1978, pp. 118-124.

48. **Mitchell, O.R.**, Target/Background Segmentation and Classification in FLIR Imagery, *Proc. of a Workshop on Image Understanding*, Cambridge, May 1978, pp. 115-117.

49. **Mitchell, O.R.** and **Lutton, S.M.**, Segmentation and Classification of Targets in FLIR Imagery, *Proc. of a Workshop on Image Understanding*, Pittsburgh, Nov. 1978, pp. 59-65.

50. **Nagao, M., Tanabe, H.** and **Ito, K.**, Agricultural Land Use Classification of Aerial Photographs by Histogram Similarity Method, *Proc. of 3rd IJCPR*, 1976, pp. 669-672.

51. **Nagao, M.** and **Tanabe, H.**, Some Trials of Signature Extension in the Analysis of Aerial Photographs, *US-Japan Seminar on Image Processing in Remote Sensing*, Nov. 1976, College Park, MD, USA.

52. **Nagy, G.**, Digital Image-Processing Activities in Remote Sensing for Earth Resources, *Proc. of IEEE*, Vol. **60**, No. 10, 1972, pp. 1170-1200.

53. **Nagy, G.** and **Tolaba, J.**, Nonsupervised Crop Classification Through Airborn Multispectral Observations, *IBM J. Res. Develop.*, Vol. **16**, No.2, 1972.

54. **Nevatia, R.** and **Binford, T.O.**, Description and Recognition of Curved Objects, *Artificial Intelligence*, Vol. **8.**, 1977, pp. 77-98.

55. **Otsu, N.**, A Threshold Selection Method from Gray-Level Histograms, *IEEE Trans. on Systems, Man and Cybernetics*, Vol. **SMC-9**, No. 1, 1979, pp. 62-66.

56. **Pavlidis, T.** and **Horowitz, S.L.**, Picture Segmentation by a Directed Split-and-Merge Procedure, *Proc. of 2nd IJCPR*, Aug. 1974.

57. **Pavlidis, T.**, A Review of Algorithms for Shape Analysis, *Computer Graphics and Image Processing*, Vol. **7**, 1978, pp. 243-258.

58. **Pratt, W.K.**, *Median Filtering*, USCIPI Report 620, Univ. of Southern Calif., 1975, pp. 116-123.

59. **Pratt, W.K.**, *Digital Image Processing*, John Wiley & Sons, 1978.

60. Quam, L.H., Road Tracking and Anomaly Detection in Aerial Imagery, *Proc. of a Workshop on Image Understanding*, Cambridge, May 1978, pp. 51-55.

61. Riseman, E.M. and **Hanson, A.R.**, *Design of a Semantically-Directed Vision Processor*, Tech. Rept. 75C-1, Dept. of Computer and Information Science, Univ. of Mass., 1975.

62. Riseman, E.M. and **Arbib, M.A.**, Computational Techniques in the Visual Segmentation of Static Scenes, *Computer Graphics and Image Processing*, Vol. **3**, 1977, pp. 221-276.

63. Rosenfeld, A., Compact Figures in Digital Pictures, *IEEE Trans. on Systems, Man and Cybernetics*, Vol. **SMC-4**, 1974, pp. 221-223.

64. Rosenfeld, A. and **Kak, A.C.**, *Digital Picture Processing*, Academic Press, 1976.

65. Rosenthal, D. and **Bajcsy, R.**, Conceptual and Visual Focussing in the Recognition Process as Induced by Queries, *Proc. of 4th IJCPR*, 1978, pp. 417-420.

66. Rubin, S.M., The ARGOS Image Understanding System, *Proc. of a Workshop on Image Understanding*, Pittsburgh, Nov. 1978, pp. 159-162.

67. Russell, D.M. and **Brown, C.M.**, Representing and Using Locational Constraints in Aerial Imagery, *Proc. of a Workshop on Image Understanding*, Pittsburgh, Nov. 1978, pp. 152-158.

68. Sakai, T., **Kanade, T.** and **Ohta, Y.**, Model-Based Interpret. of Outdoor Scene, *Proc. of 3rd IJCPR*, 1976, pp. 581-585.

69. Shirai, Y., A Context Sensitive Line Finder for Recognition of Polyhedra, *Artificial Intelligence*, Vol. 4, 1973, pp. 95-119.

70. Shortliffe, E., *Computer-Based Medical Consultations: MYCIN*, American Elsevier, New York, 1976.

71. Steven, A.L., **Zucker, W.** and **Rosenfeld, A.**, Iterative Enhancement of Noisy Images, *IEEE Trans. on Systems, Man and Cybernetics*, Vol. **SMC-7**, 1977, pp. 435-442.

72. Swain, P.H. and **Davis, S.M.**, eds., *Remote Sensing: The Quantitative Approach*, McGraw-Hill, New York, 1978.

73. Tamura, H., **Mori, S.** and **Yamawaki, T.**, Effectiveness of Textural Features for Classification of Aerial Multispectral Images, *Proc. of IEEE Conf. on Pattern Recognition and Image Processing*, 1977, pp. 289-298.

74. Tanimoto, S.L. and **Pavlidis, T.**, A Hierarchical Data Structure for Picture Processing, *Computer Graphics and Image Processing*, Vol. 4, 1975, pp. 104-119.

75. Tenenbaum, J.M. and **Barrow, H.G.**, *Experiments in Interpretation-Guided Segmentation*, SRI Tech. Note 123, 1976.

76. Tomita, F. and **Tsuji**, S., Extraction of Multiple Regions by Smoothing in Selected Neighborhoods, *IEEE Trans. on Systems, Man and Cybernetics*, Vol. **SMC-7**, 1977, pp. 107-109.

77. Tomita, F., **Shirai**, Y. and **Tsuji**, S., Classification of Textures by a Structural Analysis, *Proc. of 4th IJCPR*, 1978, pp. 556-568.

78. Tsuji, S. and **Tomita**, F., A Structural Analyzer for a Class of Textures, *Computer Graphics and Image Processing*, Vol. **2**, 1973, pp. 216-231.

79. Uhr, L., Layered "Recognition Cone" Networks that Preprocess, Classify and Describe, *IEEE Trans. on Comput.*, Vol. **C-21**, 1972, pp. 758-768.

80. Waltz, D., *Generating Semantic Descriptions from Drawings of Scenes with Shadows*, Ph.D. Thesis, M.I.T., 1972.

81. Weszka, J.S., **Dyer**, C.R. and **Rosenfeld**, A., A Comparative Study of Texture Measures for Terrain Classification, *IEEE Trans. on Systems, Man and Cybernetics*, Vol. **SMC-6**, 1976, pp. 269-285.

82. Weszka, J.S., A Survey of Threshold Selection Techniques, *Computer Graphics and Image Processing*, Vol. **7**, 1978, pp. 259-265.

83. Winston, P.H., ed., *The Psychology of Computer Vision*, McGraw-Hill, New York, 1975.

84. Yakimovsky, Y. and **Feldman**, J.A., A Semantics Based Decision Theoretic Region Analysis, *Proc. of 3rd IJCAI*, 1973.

85. Zucker, S.W., **Rosenfeld**, A. and **Davis**, L.S., General Purpose Models: Expectations about the Unexpected, *Proc. of 4th IJCAI*, 1975, pp. 716-721.

86. Zucker, S.W., Toward a Model of Texture, *Computer Graphics and Image Processing*, Vol. **5**, 1976, pp. 190-202.

APPENDIX

Overview of the Authors' Laboratory

The laboratory was established in 1973 at the Department of Electrical Engineering, Kyoto University for the purpose of investigating the models of human intellectual activities, and developing their application systems. Our activities belong to the study of artificial intelligence, but are always based on engineering approaches.

The main research topics so far have been the following:

1. **Pattern Recognition and Picture Processing**

    (1)   development of image processing software system

    (2)   edge-preserving smoothing

    (3)   structural approach to texture analysis

    (4)   remote sensing image analysis

    (5)   structural analysis of aerial photographs

    (6)   application of production system concept to the analysis of aerial photographs

    (7)   line-drawing processing and recognition

    (8)   map information processing and its data base system

2. **Mechanical Processing of Natural Languages**

    (1)   a programming language for the analysis and synthesis of sentences

    (2)   a programming language incorporating Chinese characters

    (3)   text editor for Japanese language

    (4)   morphological analysis of Japanese language

    (5)   segmentation of long compound words of Japanese

    (6)   syntactic and semantic analysis of Japanese sentences

    (7)   contextual analysis of Japanese sentences

(8) mechanical translation of titles of English scientific and technical papers into Japanese

(9) mechanical translation of Japanese programming manuals into English

(10) computer utilization of ordinary Japanese and English dictionaries

(11) keyword extraction and automatic indexing

3. Special Digital Systems for Information Processing

(1) design and construction of a LISP machine

(2) design and construction of an image processing system

(3) line-drawing input-and-correct system using a β-dimensional graphic display device

(4) I/O terminal of Japanese language

4. Artificial Intelligence and Learning Systems

(1) semantic network and knowledge representation

(2) question-answering system

(3) dialogue system

The information processing facilities shown in Fig. 1 were constructed in the laboratory. Particularly, the interactive color display system for image analysis, and the LISP machine NK-3 for language processing were new designs at the time of the development, and had several creative features.

Besides these facilities, six TSS terminals including a Chinese character I/O terminal are available in the laboratory. They are connected to a very big computer system FACOM M200 at the computer center of Kyoto University. Large sophisticated programs of the above research topics have been developed on the computer system at the computer center.

The laboratory is run under the direction of Professor Makoto Nagao. The main staff members are Associate Professor Jun-ichi Tsujii, who is specialized in language processing, and Assistant Professor Takashi Matsuyama, who is specialized in image processing. About fifteen graduate and undergraduate students are conducting studies in various research topics.

Prof. Makoto Nagao
Department of Electrical Engineering
Kyoto University
Sakyo, Kyoto, 606
Japan

COMPUTER FACILITIES

The schematic drawing of the computer facilities in our laboratory is shown in Fig. 1. Three minicomputers, TOSBAC-40A, TOSBAC-40C, and INTERDATA 8/32, extend their private I/O buses to the switching box, by which we can assign such devices as LP, MT, TV, etc., to any one of the three computers.

The LISP machine NK-3 shares the main memory (512K byte) with INTERDATA 8/32. All I/O instructions from NK-3 are simulated by the I/O programs of INTERDATA 8/32. Therefore, NK-3 can use any devices belonging to INTERDATA 8/32 without installing special hardware interfaces.

The large scale magnetic disc system (100M byte and 200M byte disc drives) is installed for the storage of large data such as LANDSAT image data for image processing and various large dictionaries for natural language processing. Although this disc system is at present connected only to TOSBAC-40C, it will be dual accessible from INTERDATA 8/32 in the near future.

The following are the specifications of the main devices:

(1) **INTERDATA 8/32**
     32 bit machine
     512K byte core memory (maximum 1M byte)
     Effective cycle time: 0.3μs
     User microprogrammable (512 words Writable Control Store)

(2) **TOSBAC-40C**
     16 bit machine
     64K byte core memory
     Memory cycle: 0.8μs
     Fast Fourier Transform (FFT) firmware

(3) **Bulk Core Memory**
     512K byte (maximum 1M byte)
     Transfer rate: about 900KB/s

(4) **Large Scale Magnetic Discs**
     Capacity   : 100M byte, 200M byte
     Transfer rate: 806KB/s

(5) **Drum Type Film Scanner**
     Sampling pitch: 50μm, 100μm, 200μm
     Spot size   : 50μm
     Film size   : maximum 15 × 15 cm
     Scanning   : Red, Green, Blue, B/W, 8 bits each,
                     simultaneously
     Light source : halogen lamp (R.G.B.), He-Ne laser (B/W)

(6) **Flying Spot Scanner (FSS)**
     Addressable points: 4,096 × 4,096
     Spot size   : 70μm
     Mainly used as color film recorder

(7) **Interactive Color Image Display System**
Resolution: 320 × 250
Two black/white image outputs with 6 bits for each image.
Output with 4 bits for each R, G and B.
Color-image output with 4 bits for each R, G and B.
Input by joystick line drawing.

(8) **Point & Line Drawing Tablet**
Resolution: 22,000 × 22,000
Area   : 22 × 22 inch
Mainly used for input of map information.

(9) **Three-Dimensional Graphic Display**
Resolution   : 4,096 × 4,096
Intensity levels: 32
Image scaling, translation, rotation and three dimensional
matrix multiplication (by hardware). Interactive display
by keyboard and light pen.

(10) **Matrix Printer Plotter (MPP)**
Mainly used for quick look recording of pictures and
images, and for the text-editor printer.

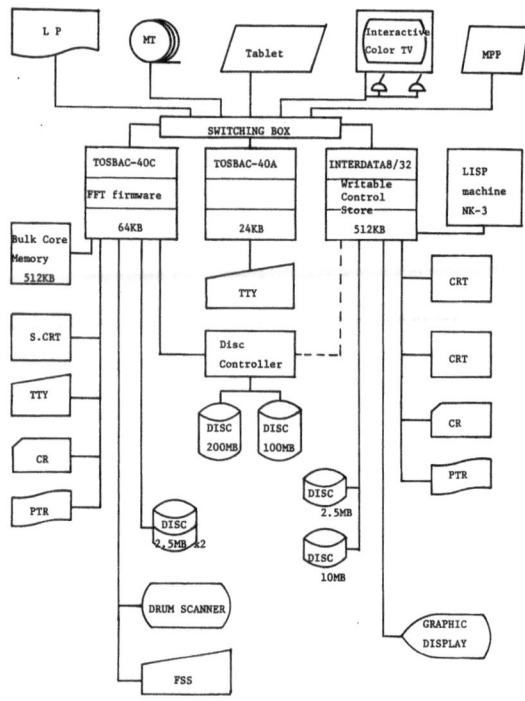

INDEX